LabVIEW - A Flexible Environment for Modeling and Daily Laboratory Use

Edited by Riccardo de Asmundis

Published in London, United Kingdom

IntechOpen

Supporting open minds since 2005

LabVIEW - A Flexible Environment for Modeling and Daily Laboratory Use
http://dx.doi.org/10.5772/intechopen.91583
Edited by Riccardo de Asmundis

Contributors
Muhammad Asraf Hairuddin, Nur Dalila Khirul Ashar, Amar Faiz Zainal Abidin, Nooritawati Md Tahir, ahmad farid abidin, Mohd Abdul Talib Mat Yusoh, R.B. Malathy, Govardhan Bhat, U.K. Dewangan, Janani Rajaraman, Andrea Rubano, Rohit Kumar, Qiucheng Yu, Domenico Paparo, Mehmet Ertugrul, Ahmet Mavi, Ahmet Özmen, Prema Ramasamy, Shri Tharanyaa, Abin Sathesan, Trinh Quang Duc, Giuseppe Porzio

First published in London, United Kingdom, 2021 by IntechOpen
IntechOpen is the global imprint of INTECHOPEN LIMITED, registered in England and Wales, registration number: 11086078, 5 Princes Gate Court, London, SW7 2QJ, United Kingdom
Printed in Croatia

British Library Cataloguing-in-Publication Data
A catalogue record for this book is available from the British Library

Additional hard and PDF copies can be obtained from orders@intechopen.com

LabVIEW - A Flexible Environment for Modeling and Daily Laboratory Use
Edited by Riccardo de Asmundis
p. cm.
Print ISBN 978-1-83968-840-9
Online ISBN 978-1-83968-841-6
eBook (PDF) ISBN 978-1-83968-842-3

For EU product safety concerns:
IN TECH d.o.o., Prolaz Marije Krucifikse Kozulić 3, 51000 Rijeka, Croatia, info@intechopen.com or visit our website at intechopen.com.

Meet the editor

Riccardo de Asmundis is a physicist who graduated cum laude from the University of Naples "Federico II" in 1987, where he is also a contract professor teaching the Electronics Fundamentals course. He is a senior researcher at Istituto Nazionale di Fisica Nucleare (INFN) in Naples and a Certified LabVIEW Developer (CLD) and Professional Instructor (CPI) for National Instruments. He is a member of international collaborations in high-energy physics, currently for the ATLAS and SND Experiments at CERN; he spent several decades abroad, designing particle detectors and data acquisition systems. Dr. de Asmundis has authored more than 1100 publications in international reviews on topics such as high-energy physics, particle detectors, and photon detectors. He owns his private electronic lab where he renovates electronic equipment for music, such as keyboards, synthesizers, organs, radios, and HiFi systems. He also plays classical piano.

Contents

Preface

The LabVIEW software environment from National Instruments (NI) is a reference used by engineers and scientists worldwide. It is known for being used as a specific solution and, normally, is coupled with NI-made or third-party hardware or in conjunction with additional software. There is no laboratory or industry that does not adopt LabVIEW as a standard.

Since its inception in 1986, LabVIEW has grown enormously. It provides not only an annual upgrading policy, but also a variety of annexed software products such as supporting software and design tools for measurements, automation, simulation, hardware integration, and data analysis.

LabVIEW is wrongly regarded as a simple tool for acquiring, processing, and displaying data. On the contrary, it is an extremely powerful and complete programming language. The peculiarity of its graphical development interface, implemented in the so-called G-programming language (where G stands for "graphical"), is both an advantage and disadvantage of LabVIEW. Its implicitly simple approach based on graphical diagrams could deceive the user or potential developers, inducing them to think that LabVIEW is a simple or even a trivial way of programming basic or simple tasks. However, in spite of its intuitive interface, LabVIEW needs to be carefully understood and its development techniques must be acquired and well known in order to develop professional applications that are robust, readable, scalable, and maintainable. The knowledge of the language elements is far enough from the ability to develop good and effective applications. With the aim to accomplish that, it is suggested that beginners consistently study and practice using LabVIEW, while advanced users should keep abreast of updates to the software. Courses, webinars, documentation, forums, discussions, experience, and formation, in general, should be considered as a part of the personal patrimony as a professional LabVIEW developer. Moreover, developers should think about good and specific planning before they start to merely "design," in the sense of trying to put elements together to get a result.

To guarantee a complete and effective knowledge of the product, users should take into account the certification program available from NI. This program is a graduation process divided into three levels and based on exams. LabVIEW certification needs to be renewed every three years, making the owner responsible for keeping up with updates in the field. This is particularly important, especially since NI introduced a new platform for LabVIEW called LabVIEW NXG (New Generation), which should take the place of the "old" (but currently still used) traditional LabVIEW platform.

Given all of this, it is clear that LabVIEW has applications in many fields. As such, the chapters in this book cover such topics as didactics, laboratory applications of tests and measurements, and computational and modeling applications.

The book is organized into two sections: "LabVIEW as a Laboratory Tool" and "LabVIEW in Modeling." The first section contains six chapters and the second section contains three chapters.

Chapter 1, "Analyzing and Presenting Data with LabVIEW," may appear to fit better in the second section of the book. However, I chose it as an introductory chapter because it contains a detailed summary of the calculation and analysis functions present in LabVIEW, giving the reader a wide overview of the possibilities that the development environment provides, which new users of the system often overlook. The same can be said about the aspects of data presentation made with LabVIEW, which is a peculiarity often delegated to further "external" software, an action that is not always necessary.

Chapter 2, "TeraVision: A LabVIEW Software for THz Hyper-Raman Spectroscopy," is a review of a complete research lab application carried out with LabVIEW. The chapter opens with a premise on terahertz physics, necessary to understand the nature of the application, and then moves on to the description of the measurement implementation using LabVIEW. It is interesting to note the complexity of the application, mainly due to the considerable quantity of instruments and elements to be controlled centrally.

Chapters 3, 4, and 5 deal with the same theme: everyone is looking for, and finding, a way to interface LabVIEW software with lean and affordable hardware like Arduino's Open Source. Although accompanied by respectable hardware from NI, LabVIEW can also be used easily with third-party hardware. This solution can be used where there is no need for stringent and absolute reliability (typical of industrial applications) but one can be satisfied with cheaper solutions. In this field, Arduino represents a frontier that has always been widely considered, given its low cost compared to that of NI hardware. Chapter 3 opens with a discussion of the management, economic and otherwise, of resources for an educational laboratory, with an eye toward the "remote control" of laboratories. Chapter 4 instead points directly to data acquisition techniques, with a focus on the concrete aspects of the electrical type. Chapter 5 deals with the practical aspects of interfacing with Arduino, presenting two different technical approaches to the integration between the parts and focusing on different platforms such as Raspberry.

Chapter 6, "Certain Applications of LabVIEW in the Field of Electronics and Communication", generically deals with the applications of signal treatments that can be carried out in LabVIEW in various fields. These include pattern recognition to use with embedded systems as well as applications in robotics, medicine, and education.

The second section begins with Chapter 7, which presents an algorithm, built entirely in LabVIEW, for determining the structural response of buildings to strong stimuli such as earthquakes. The chapter also presents some historical earthquakes that have been studied and analyzed using LabVIEW, using a discrete-time integration computation model. It concludes with a positive opinion regarding the use of LabVIEW for this application.

Chapter 8, "LabVIEW as Power Disturbances Classification Tools," presents an interesting application of LabVIEW in the field of monitoring the quality of electrical networks for power distribution. The system is based on a hybrid architecture, which uses standard commercial electrotechnical tools placed in the field and ends with an interface to LabVIEW where the data is collected and subsequently analyzed.

Chapter 9, "Digital System Design," uses LabVIEW for modeling digital logic. This represents a useful opportunity in the field of teaching, where potential learners can build combinatorial logic based exclusively on gates and elementary logic. This activity represents a strong stimulus to the understanding of the realization of advanced logic networks, such as those built-in small- and medium-scale integrated circuits starting from basic logic.

LabVIEW offers endless possibilities and, as we said at the beginning, it must be imagined as a programming language in all respects. The use of LabVIEW requires passion, method, and rigor, which are all essential skills to obtain professional results. In particular, it should be pointed out that LabVIEW must be approached methodically, avoiding improvisations and attempts supported by the fact that the use of the G-language is implicitly easy; starting from this assumption would represent a false step. Any professional use of LabVIEW must be supported by adequate preparation, starting with the execution of the appropriate existing basic courses (Core 1, Core 2, and Core 3) and proceeding hand in hand with personal experiences, towards any advanced courses.

Remember, the G is not just for graphics; the G combined with the correct programming structures and strategies makes a LabVIEW program worthy of the name.

Riccardo de Asmundis
INFN (Istituto Nazionale di fisica nucleare),
Section of NAPOLI, Italy

Università di Napoli "Federico II",
Department of Physics,
Napoli, Italy

Section 1

LabVIEW as a Laboratory Tool

Chapter 1

Analyzing and Presenting Data with LabVIEW

Ahmet Mavi, Ahmet Özmen and Mehmet Ertuğrul

Abstract

LabVIEW is an abbreviation for Laboratory Virtual Instrument Engineering Workbench and allows scientists and engineers to develop and implement an interactive program. LabVIEW has been specially developed to take measurements, analyze data, and present the results to the user. You determine what the device looks like, rather than the manufacturer of the device. LabVIEW has a very large library of functions and subprograms (subVIs) that can help you during your programming and use without occupying memory. Hidden programming problems that you may encounter in traditional programming languages are less common in LabVIEW. LabVIEW also includes different applications such as serial device control, data analysis, data presentation, data storage and communication over the internet. Analysis library; It includes versatile and useful functions such as signal generation, signal processing filters, Windows statistics and regressions, linear algebra and array arithmetic. Due to the graphical nature of LabVIEW, it is an innate data presentation package. You can view the data in any form you want. Chart, graph and user-defined graph are among the output options that can be used. As a scientist or an engineer, you frequently measure physical changes such as temperature, pressure, time, mass, electric current, light intensity, radioactivity etc. You generally need to analyze and present the data. When you have large amounts of data, you need to use software to analyze and present the data. LabVIEW makes these actions easy for you. Because LabVIEW includes hundreds of built-in and add-on functions you need that make it easy to create a user-friendly interface. In this chapter, we focus on data analysis and presentation.

Keywords: data analysis, data presentation, report generation, DIAdem, OriginPro

1. Introduction

Almost all LabVIEW applications include 3 steps: (1) acquiring data, (2) analyzing and processing the data, and (3) presenting the data in a report or on a chart/graph (**Figure 1**).

Acquire: NI (National Instruments) is a global leader in computer-based data acquisition. More than millions of data acquisition devices have been sold by NI. LabVIEW developed by NI is a user friendly programming interface and easily communicates to NI devices. Therefore, most of the scientists and engineers choose LabVIEW for programming and NI devices for measurements.

Analyze: LabVIEW software has more than 600 built-in functions for signal synthesis, frequency analysis, probability, statistics, math, curve fitting,

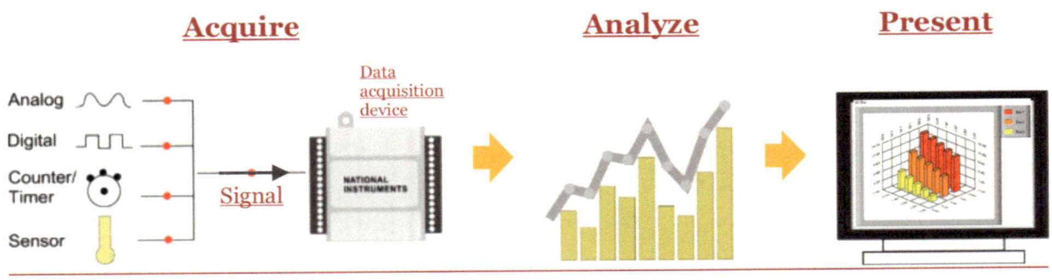

Figure 1.
Common steps in a VI.

interpolation, digital signal processing, and more. You can also reach more functions from additional modules. Unfortunately, some modules are not free.

Present: After you acquire and analyze data, you generally want to present your data. Data presentation means data visualization, report generation, data storage, Web publishing, database connectivity, data management, and more. The LabVIEW includes hundreds of functions and tools for data presentation. These allow you to visualize data in a very simple and effective manner.

2. Analyze with LabVIEW

2.1 LabVIEW data analysis and math libraries

LabVIEW presents the VIs and functions on the Functions palette using menus. When LabVIEW is installed, built-in menus appear on the Functions palette. Some of them do not contain functions by default. After installed certain modules, toolkits, and drivers, functions appear in the built-in categories. For example, Measurement I/O menu does not contain NI DAQmx by default. If you want to utilize its function, you must install NI DAQmx.

LabVIEW includes hundreds of built-in and add-ons functions for analysis. You can find a short list of analysis functions below [1].

Measurement.

Waveform-Based**.

Averaged DC-rms.
Cycle average and rms.
Pulse transition (rise, slew, overshoot).
Pulse width/period/duty.
Pulse amplitude and levels.
Signal noise and distortion (SINAD) analyzer.
Harmonic distortion analyzer.
Cross spectra Sine tone measurement.
FFT spectrum.
Frequency response function.
Power spectrum.
Power spectral density.

Array-Based.

AC and DC Estimator.
Amp and frequency estimate.
Amp and phase spectrum.
Auto power spectrum.
Cross power spectrum.
Harmonic analyzer.

Impulse response.
Network functions (avg).
Power and frequency estimate.
Power spectrum.
Scaled time-domain window.
Spectrum unit conversion.
Transfer function.
Signal Monitoring/Triggering.
Waveform-Based**.
Basic level trigger detection.
Limit testing.
Limit specification.
Limit specification by function.
Waveform peak detection.
Array-Based.
Peak detection.
Pulse parameters.
Threshold peak detector.
Signal Generation.
Windowing.
Digital Filters.

2.2 Statistics

1D, 2D, and 3D ANOVA.
Chi square distribution.
Contingency table.
erf(x) and erfc(x).
F distribution.
T distribution.
General histogram.
Histogram*.
Inverse chi square.
Distribution.
Inverse F distribution.
Inverse normal distribution.
Inverse T distribution.
Mean*.
Median*.
Mode*.
Moment about mean.
Mean squared error (MSE).
Normal distribution.
Polynomial interpolation.
Rational interpolation.
Root mean square (rms).
Spline interpolant/interpolation.
Standard deviation*.
Variance.
Signal Processing.
Autocorrelation.
Convolution.
Cross power.

Cross correlation.
Decimate.
Deconvolution.
Derivative x(t).
Fast Hilbert transform.
Fast Hartley transform.
Integral x(t).
FFT/Inverse FFT (Re + Im).
Inverse fast Hilbert transform.
Unwrap phase Y[i] = Clip{X[i]} Y[i] = X[i-n].
Curve Fitting.
Exponential fit.
General least squares linear fit.
General polynomial fit.
Linear fit.
Nonlinear Lev-Mar fit.
1D and 2D linear evaluation*.
1D and 2D polynomial evaluation*.
Numeric integration.
Polar to rect/rect to polar.
Scale 1D/2D.
Find polynomial roots.
Mathematics/Numerical Methods.
Digital Waveform Analysis.
Digital signal subset.
Invert digital signal.
Uncompress digital signal.
Compress digital signal.
Digital signal size.
Search for digital pattern.
Compare digital signals.
Append digital signals.
Append digital samples.
Digital waveform to Boolean array.
Boolean array to digital waveform.
Waveform Conditioning.
*Denotes VIs that are shipped with the Base package of LabVIEW for Windows.
**Waveform VIs input a time-domain signal and output a scaled measurement.

We highly recommend that check the examples located in LabVIEW before starting to create a VI for analysis. You can access the examples from NI Example Finder (**Help>> Find Examples…**). You can search the examples with keyword(s) in NI Example Finder. It also allows you to visit ni.com for more examples.

We also recommend that if possible you should choose Express VIs for analysis. An Express VI is a VI, which can be configured interactively through a dialog box. Express VIs are user friendly. You can easily configure your analysis parameters. To access to the dialog box, double-click to corresponding Express VI.

For analysis, you will frequently use **Mathematics** palette and **Signal Processing** palette (**Figure 2**).

Statistics and **Histogram** Express VIs are located in **Probability & Statistics** subpalette of **Mathematics** palette (**Figure 3**).

In the following example, VI simulates a DC signal with noise (**Figure 4**). VI also generates a histogram and result of basic statistical analysis.

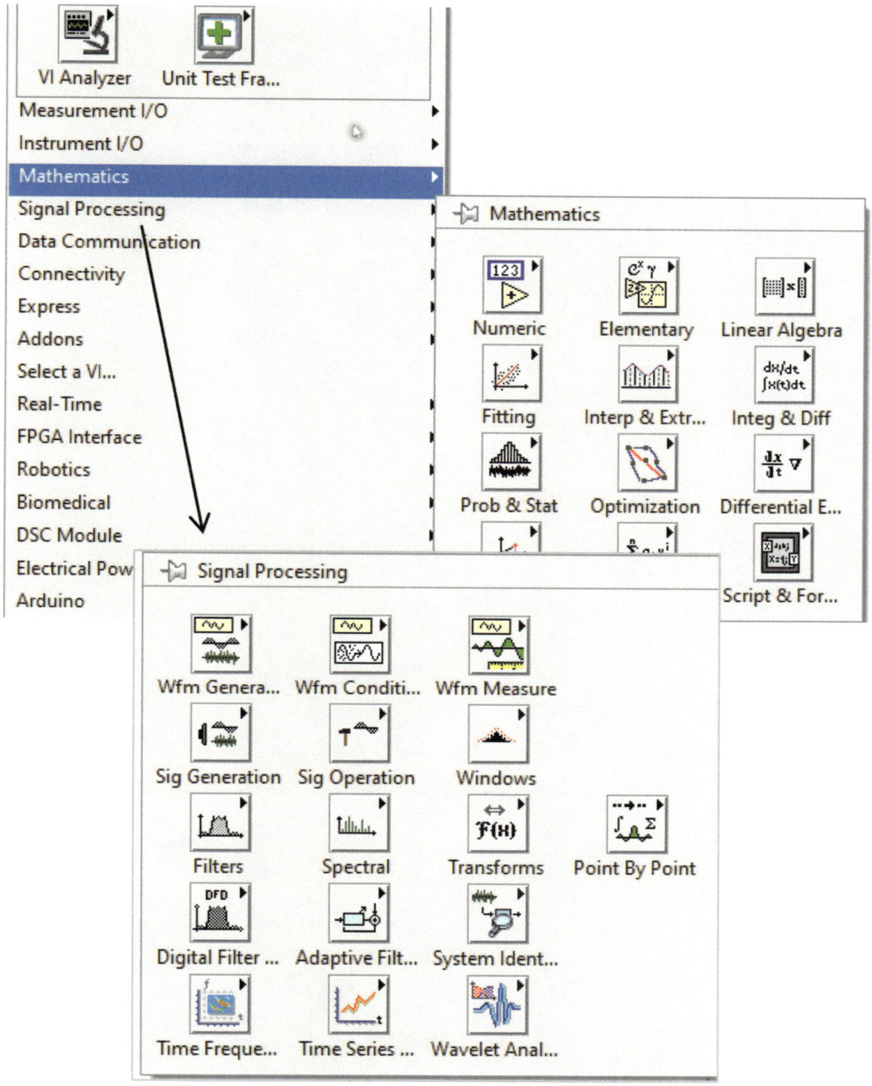

Figure 2.
Mathematics palette and signal processing palette.

Mathematics palette also contains **Fitting** subpalette (**Figure 5**). This palette contains the following fitting VIs.

- Linear Fit VI
- Exponential Fit VI
- Power Fit VI
- Gaussian Peak Fit VI
- Logarithm Fit VI

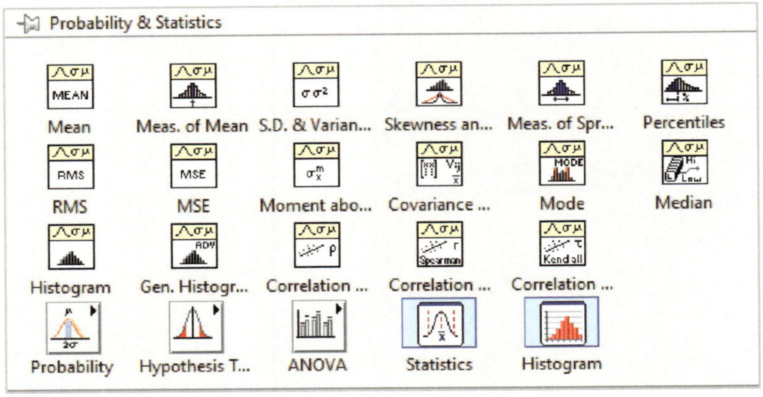

Figure 3.
Probability & Statistics palette.

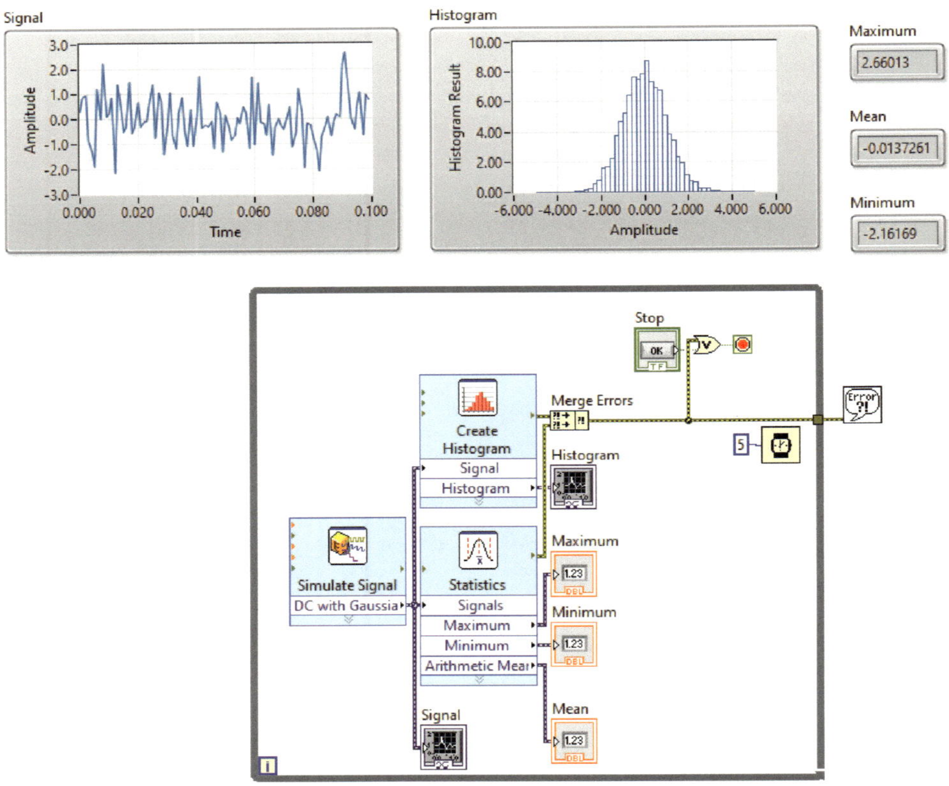

Figure 4.
Statistics and histogram express VIs in a VI.

- General Polynomial VI
- General Linear Fit VI
- Cubic Spline Fit VI
- Nonlinear Curve Fit VI

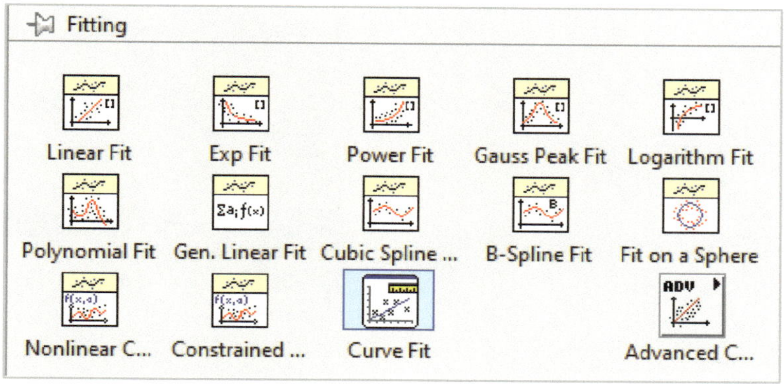

Figure 5.
Fitting VIs.

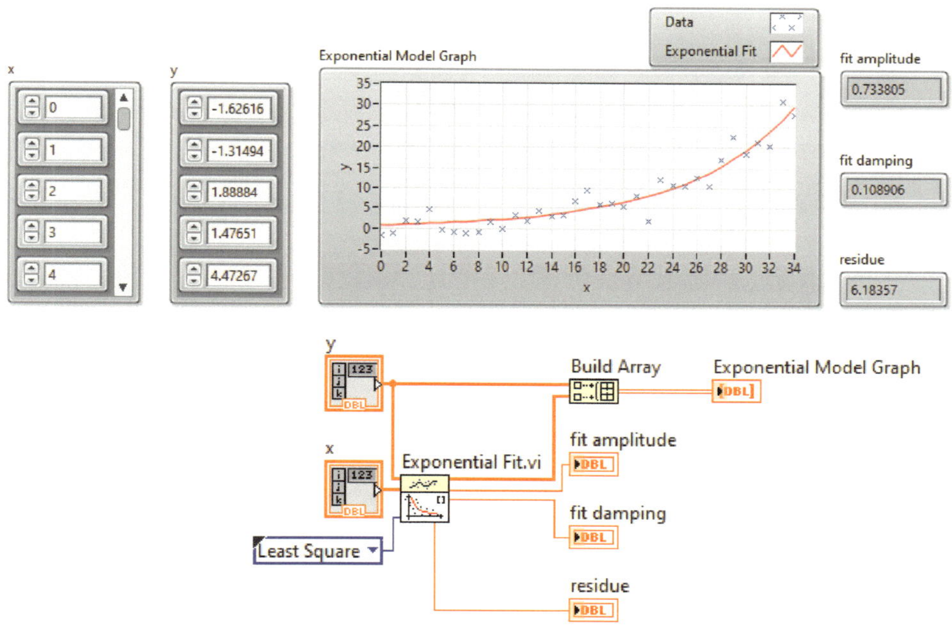

Figure 6.
Exponential fit.

You can use curve fitting for several reasons. For example, to reduce noise, to find mathematical relationships among variables, to estimate the variable value between data samples or out of data sample range.

The following simple VI plots data and Exponential fit (**Figure 6**). You can use the other fitting VIs with the same manner.

You can use the **Signal Processing** VIs for spectrum analysis, signal generation, digital filtering, and data windowing (**Figure 7**). It is located in **Functions** palette.

In **Signal Processing** palette, **Waveform Measurements** palette contains **Tone Measurements** and **Spectral Measurements** Express VIs (**Figure 8**).

The following example in **Figure 9**, **Tone Measurements** Express VI finds amplitude, frequency and phase of a signal which is generated by **Simulate Signal** Express VI. In this example, **Spectral Measurements** Express VI generates the power spectrum of the signal.

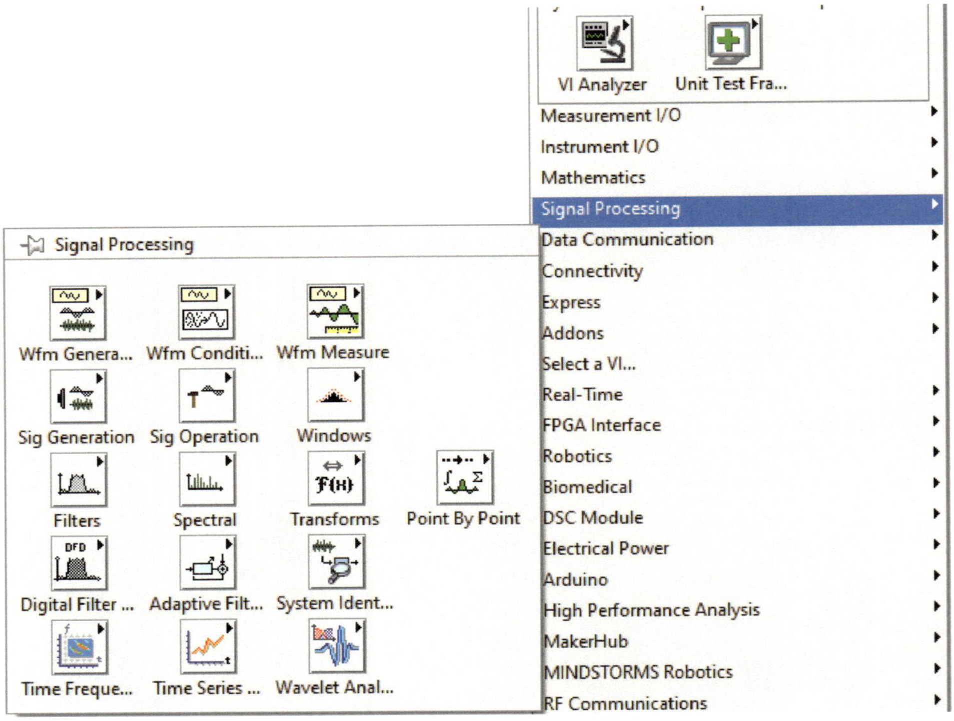

Figure 7.
Signal processing palette.

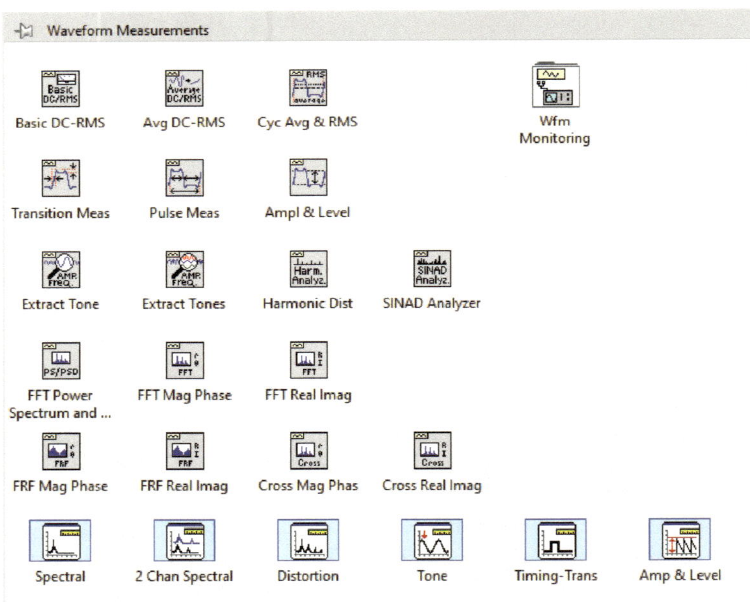

Figure 8.
Waveform measurements palette.

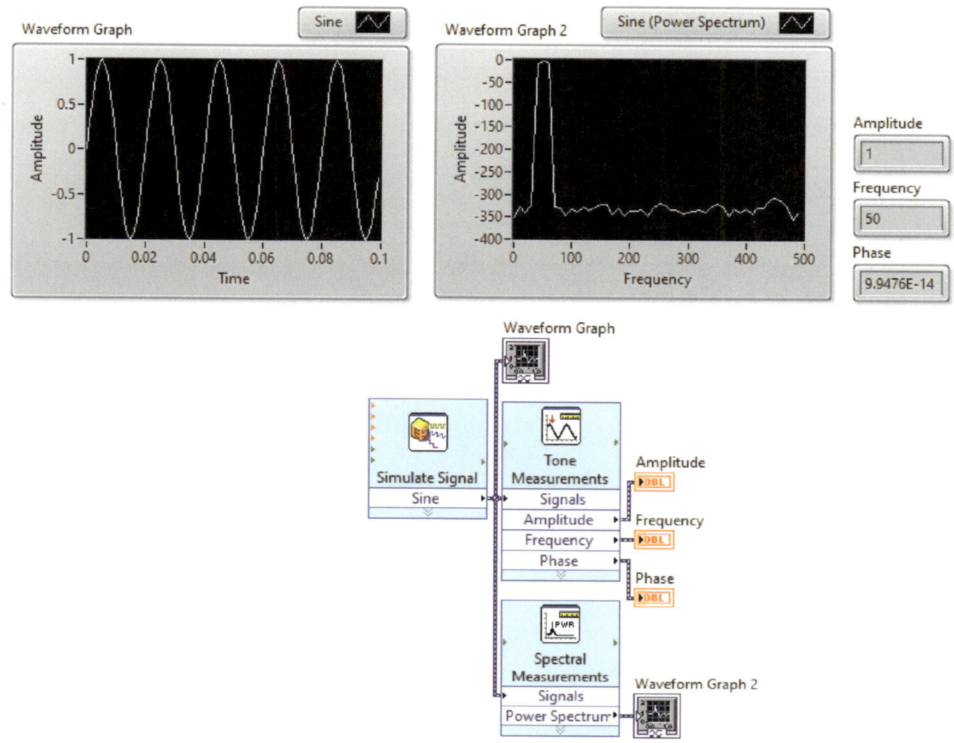

Figure 9.
Tone measurements and spectral measurements express VIs.

3. Present with LabVIEW

3.1 Visualize your data

LabVIEW includes tools for charting and graphing, 2D and 3D visualization (**Figure 10**) [2].

LabVIEW has two ways to display data in 2D. These are Chart and Graph (**Figure 11**).

A Waveform Chart remembers and displays a certain number of points by storing them in a buffer. Waveform Chart displays received data in addition to already existing data.

A Waveform Graph accepts arrays of data in various forms, e.g. array, waveform, or dynamic data. It plots all the received points at once.

You can visualize more than one data source on a chart or graph. In the following example **DAQ Assistant** take data from two channels. You can see data from all channels on a chart as shown in **Figure 12**.

A multi-plot chart can be displayed as overlaid plot or stacked plot (**Figure 13**). To select **Stack Plots** or **Overlay Plots** right-click on the chart. **Overlay Plots** mode overlays all plots on the same y-axis. **Stack Plots** mode gives the each plot its own y-axis.

To plot y values in a chart/graph you should wire only the y array data (y values) to the Waveform Chart or Waveform Graph. LabVIEW assumes that you sample y values at regular intervals, and thus creates x values at regular intervals. If you want

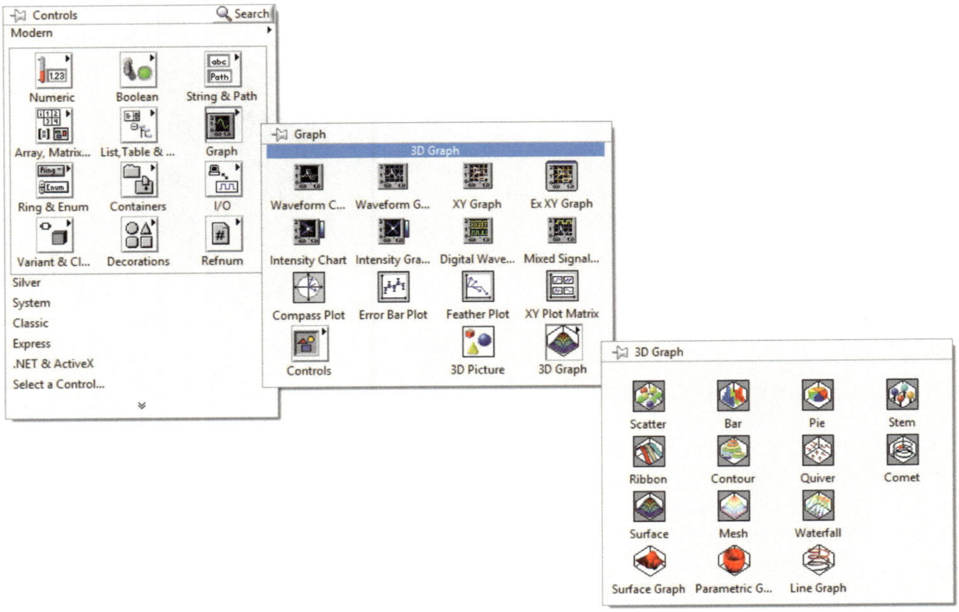

Figure 10.
2D and 3D visualization palettes.

to specify x and y values for a plot, you can use **XY Graph**. In the following example we plot multiple circles in a XY Graph (**Figure 14**).

If you want to display both analog and digital signals together in a graph, use a **Mixed Signal Graph** located in the **Graph** palette. A Mixed Signal Graph is made by bundling multiple graphable data types. You can add plot area from the pop-up menu of an existing plot area by selecting **Add Plot Area**. You can also remove a plot area by selecting **Remove Plot Area**. In the following example, you can see both analog and digital signals together in a Mixed Signal Graph with two plot area (**Figure 15**).

LabVIEW allows you to use 3D graphs to plot data in three dimensions. 3D graphs are located in **Controls> > Modern> > Graph> > 3D Graph**. LabVIEW allows you eleven types of 3D graphs: The Scatter, Bar, Pie, Stem, Ribbon, Contour, Quiver, Comet, Surface, Mesh, and Waterfall graphs. You can see some of them in **Figure 16**. However, to study with 3D graphs you must have learned basics of vector and matrix.

Plot Helper.vi is automatically created in the block diagram when you drop any of the 3D graph. **Plot Helper.vi** is a polymorphic VI and thus it can accept Matrix or Vector inputs according to your selection (**Figure 17**).

You can find the two examples of 3D graphs, below.

In the following example we created a cylinder combining 5 circles whose z axis points are different from each other. Note that **i** (iteration number) generates z matrix (**Figure 18**).

The following VI generates a sphere and visualizes it in **3D Parametric Graph** (**Figure 19**). Here radius of sphere is 5 and sphere is generated by 20 circles. A sphere is a collection of circles. You can see from XY Graph that each circles are individual size.

3.2 Publishing information to the web

LabVIEW can publish any application to the Web with Remote Panels. Therefore, you can easily make your VI reachable as a Web page. Thus, clients can control

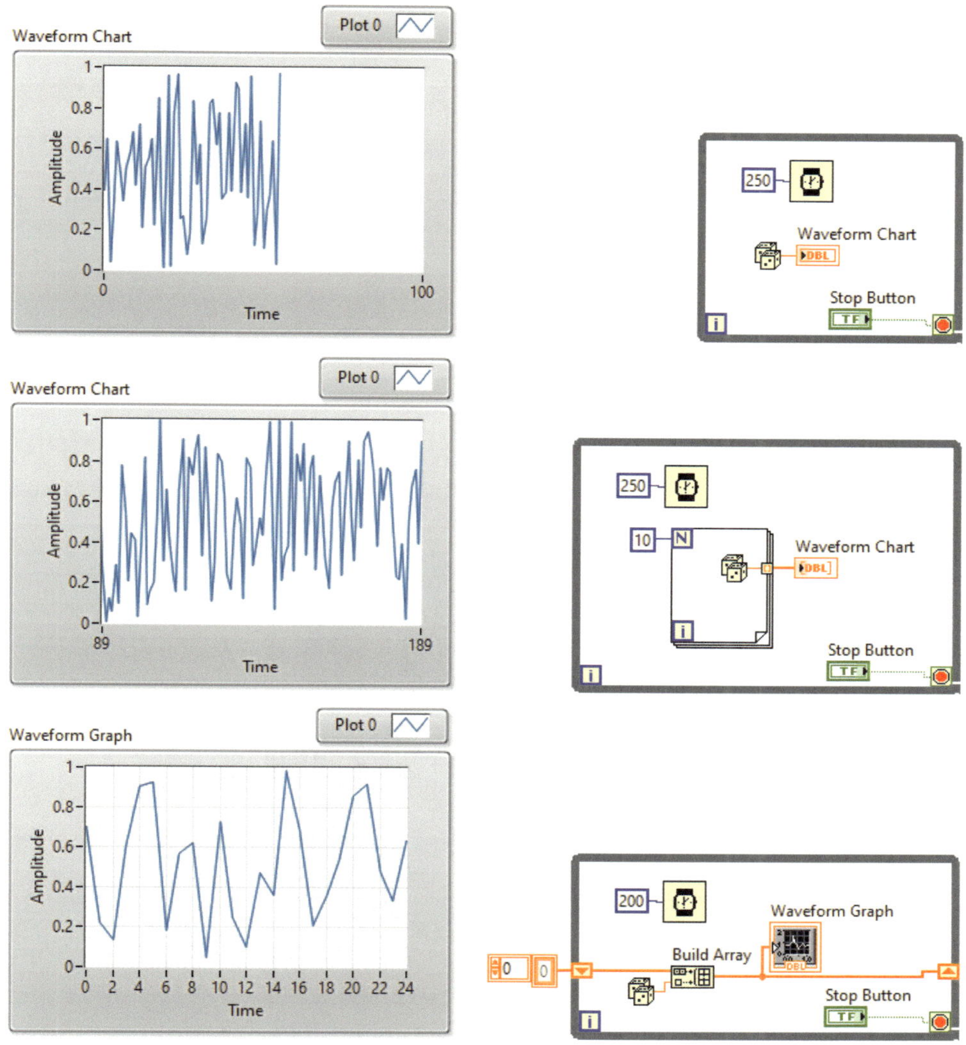

Figure 11.
Chart and graph.

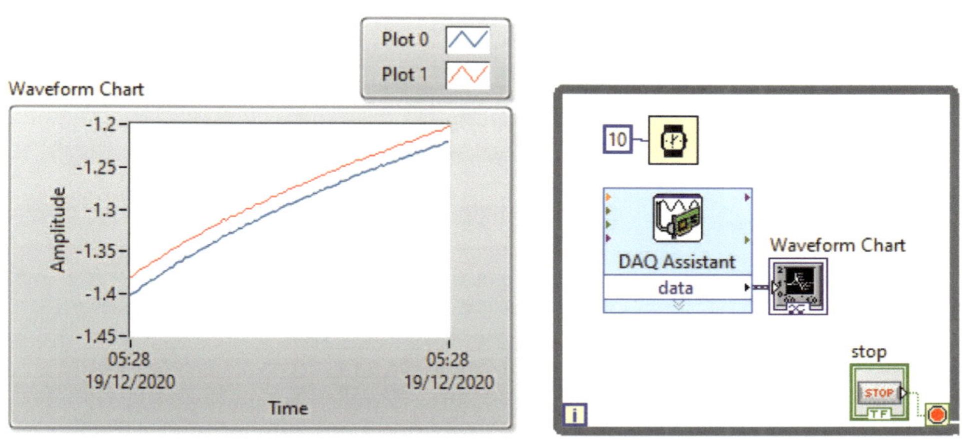

Figure 12.
Visualization of data acquired from 2 channels of a daqcard.

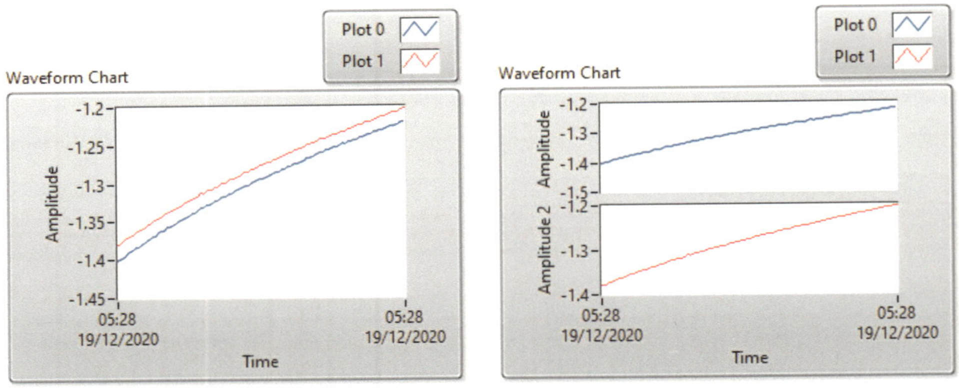

Figure 13.
Overlay plots (left) and stack plots (right) modes.

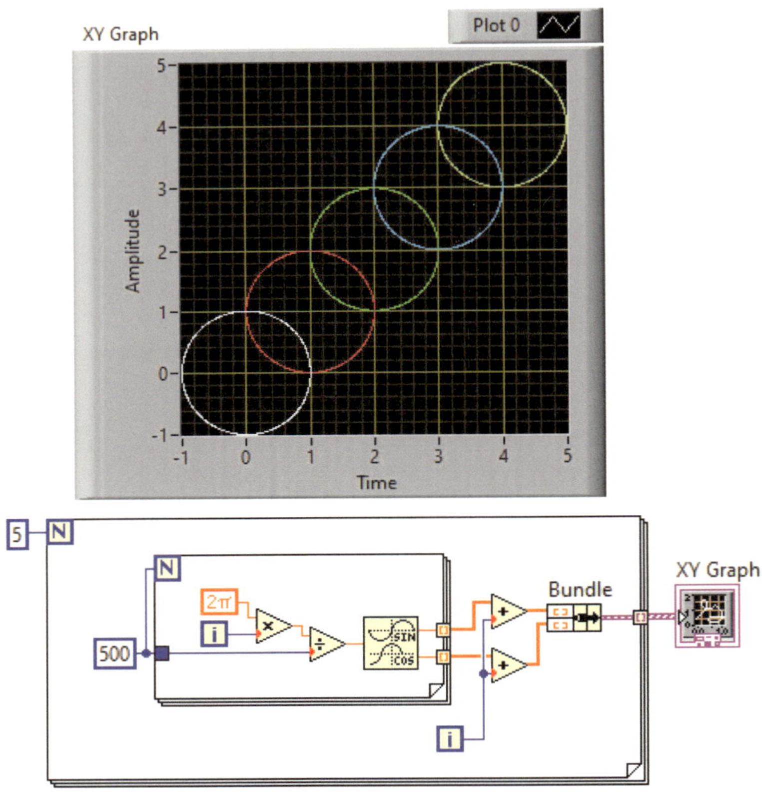

Figure 14.
Multiple plots in a XY graph.

the VI or view generated data by using their web browsers. Clients must use a version of the LabVIEW Run-Time Engine compatible with the version of LabVIEW. NI recommends that customers use the supported browser (Internet Explorer). Google Chrome version 42 and later, Mozilla Firefox 52 and later, Safari 12.1 in macOS Mojave 10.14, and Microsoft Edge are **not** supported browsers. Before view and control a front panel remotely, the Web Server must be enabled on

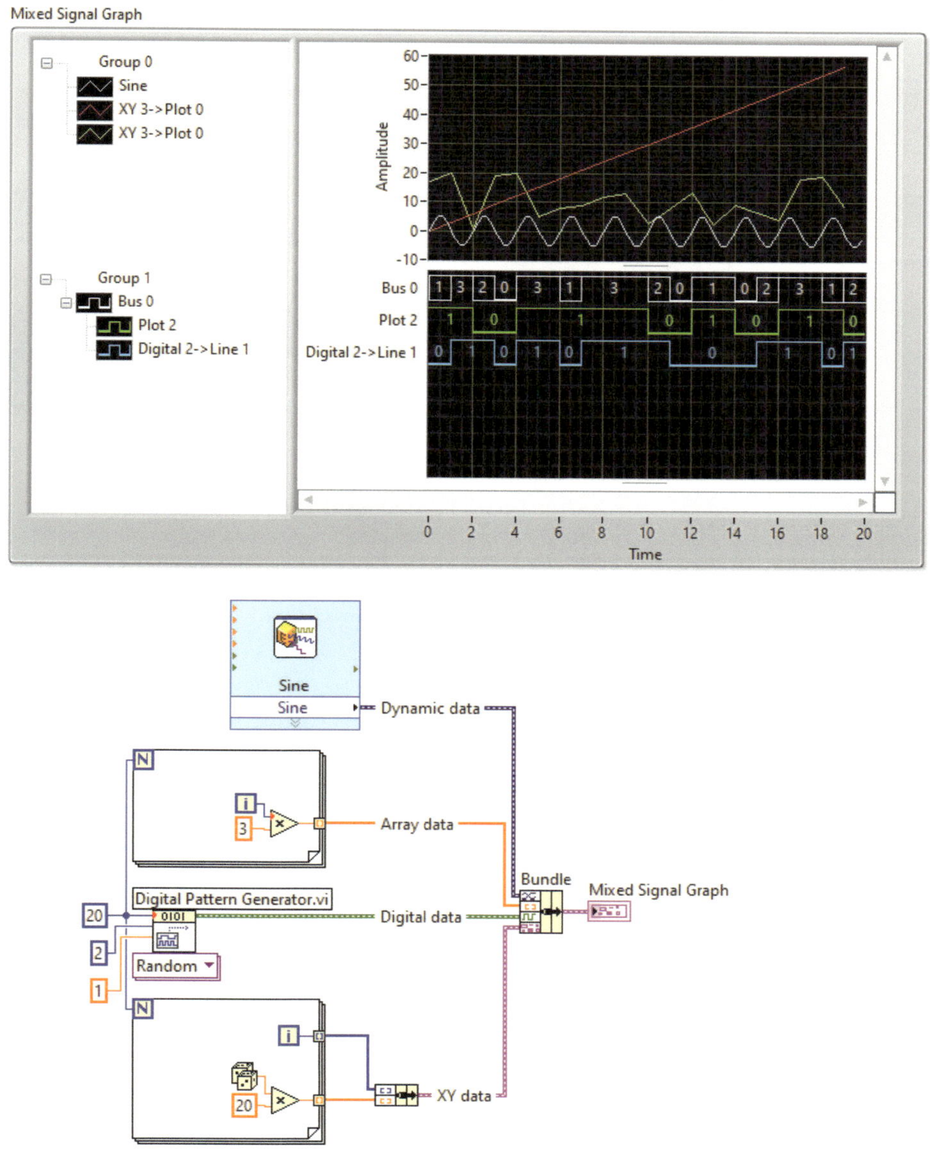

Figure 15.
Mixed signal graph with three plots.

the server computer where the VI or application wanted to view and control is located. Follow the steps below to learn how you can do it.

1. Create a VI. We created a VI named Remote Panel Example.vi (**Figure 20**).

2. Open block diagram and click to **Tools»Options»Web Server**

3. Under the Remote Panel Server section, check **Enable Remote Panel Server** (**Figure 21**).

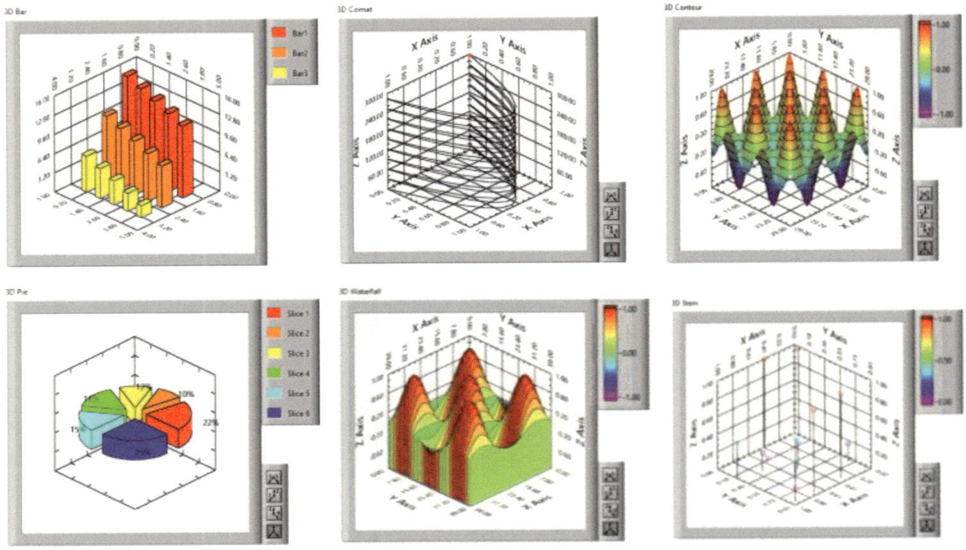

Figure 16.
3D graph examples.

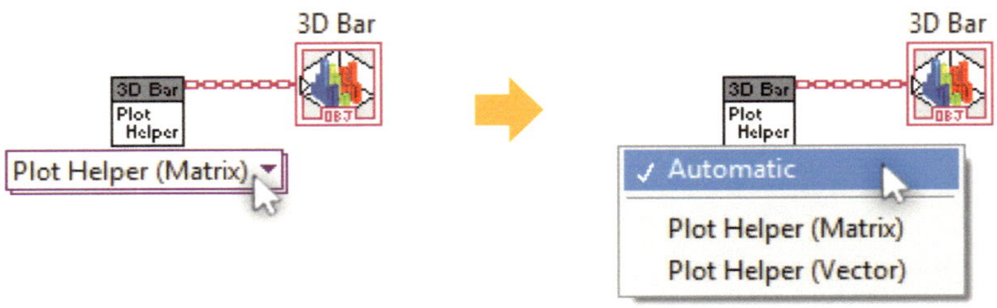

Figure 17.
Plot helper.Vi.

4. Under the **Visible VIs** section, enter the name of the VI (Remote Panel Example.vi) in the **Visible VI** field and press the **Add** button (**Figure** 22).

5. Under the **Browse Access** section, enter the network name of the computer and press the **Add** button. Allow viewing and controlling option must be selected (**Figure** 22).

6. Click OK and exit out of the **Options** dialog box.

7. Navigate to **Tools»Web Publishing Tool** to open the **Web Publishing Tool** dialog box.

8. Under **Select VI and Viewing Options** section, select the VI. After selecting appropriate **Viewing Mode** click Next (**Figure** 23).

9. Fill Document title, Header and Footer sections and click **Next** (**Figure** 24).

10. Under the **Save the New Web Page** section, select where to save the HTML file and choose the file name and press **Save to Disk** button (**Figure** 25).

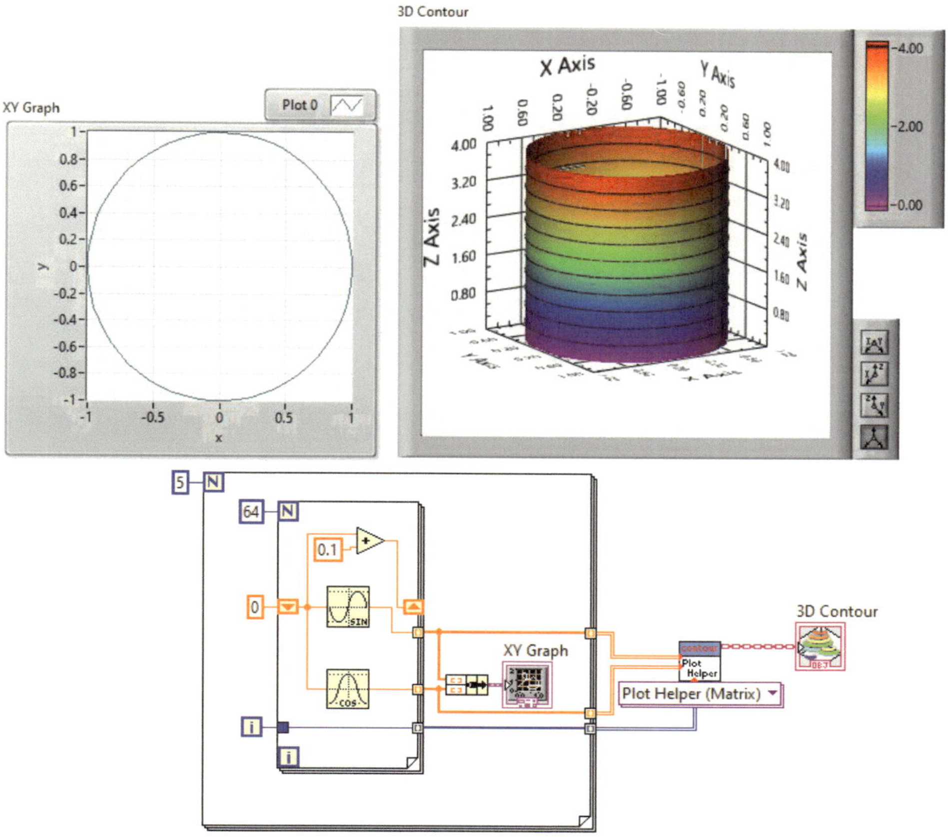

Figure 18.
Creating a cylinder in 3D counter.

11. Click **Connect** button in **Document URL** window (**Figure 26**). Ensure that default browser is Internet Explorer. If not, copy URL address and paste it to Internet Explorer.

You will see the following Internet Explorer page (**Figure 27**). In this page, clients must click Run button to control the VI from their computer.

3.3 Report generation

LabVIEW includes **Report Generation** toolkit to present your data in a Microsoft Office Word and/or Excel file. To use LabVIEW Report Generation Toolkit, it must be installed. Then corresponding functions will be located in **Functions> > Report Generation** palette (**Figure 28**).

3.3.1 Microsoft office word and excel reports

Report Generation palette contains many functions. Therefore, it is not easy to understand their properties. We recommend that you examine first the report generation example VIs in LabVIEW (**Help>> Find Examples>> Search**). You can modify them according to your purpose. These VIs generally generate reports based on templates. Using a template, allows you to generate standard reports for

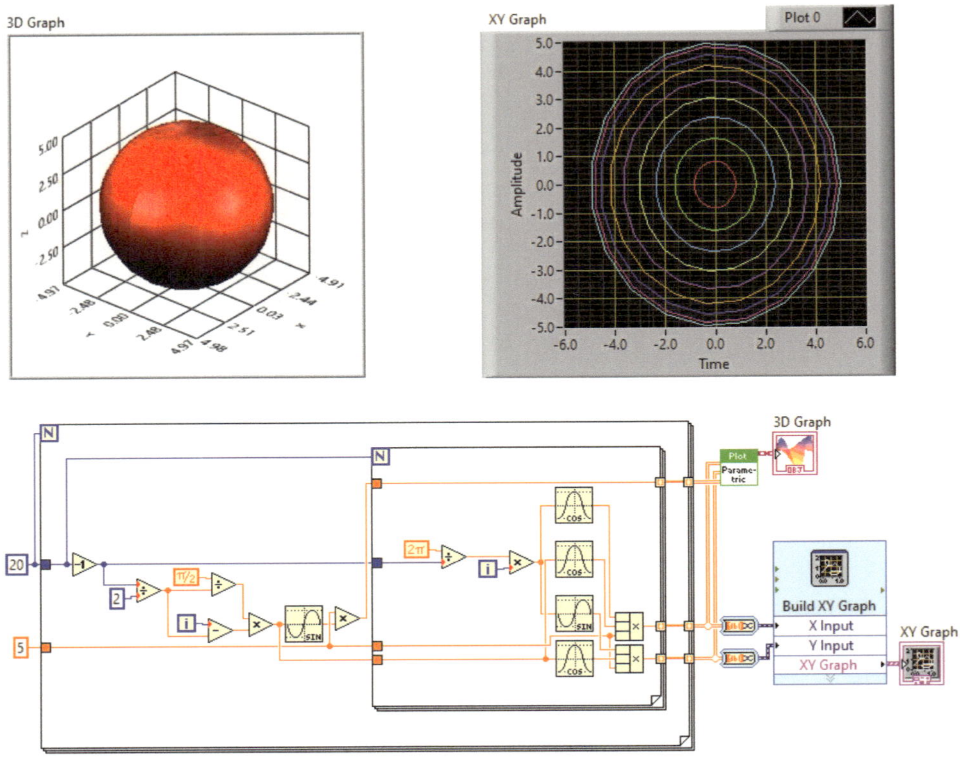

Figure 19.
Drawing a sphere.

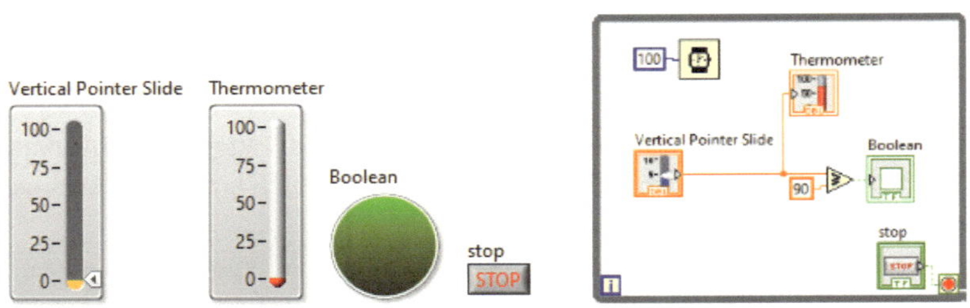

Figure 20.
Remote panel example.Vi.

each execution of a VI. In the following example we generate a report for Microsoft Office Word (**Figure 29**). The example draws a circle and paste the circle to a Ms. Office Word document (**Figure 30**). You can determine color, size, graph type, marker style etc. by using the functions in Word Specific palette (**Figure 31**). Similarly, you can use **Excel Specific** palette to generate programmatically an Excel report (**Figure 31**).

When you execute the VI you see the following picture in an automatically created Word document.

3.3.2 HTML report

LabVIEW has the ability to programmatically create html reports. Html files can be read by web browsers. We highly recommend you to present your data as a html

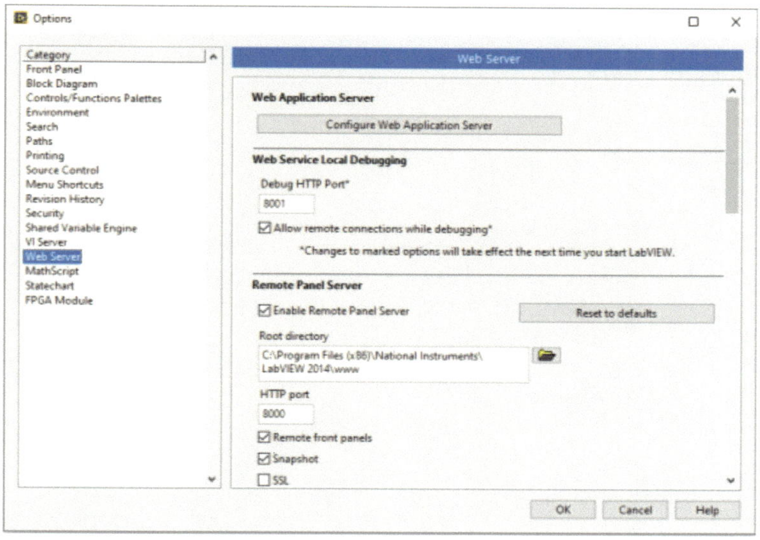

Figure 21.
Options»web server.

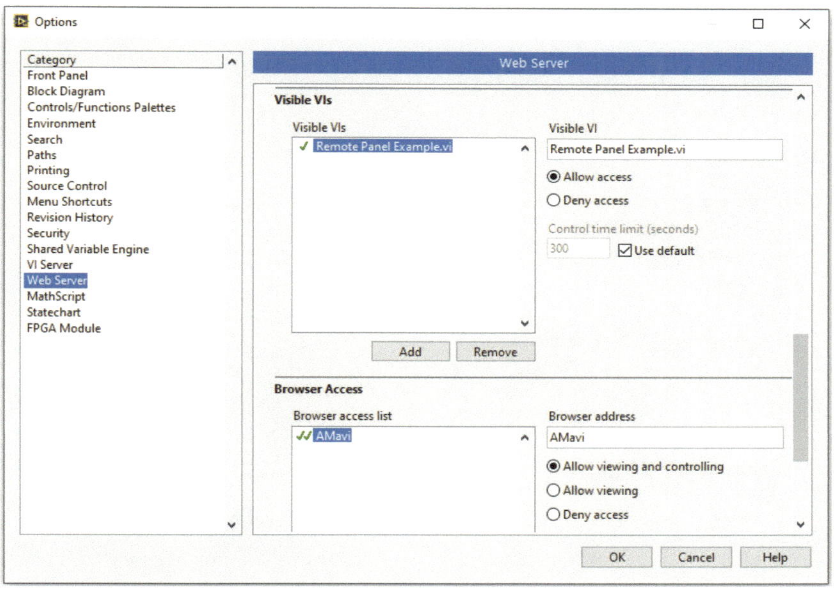

Figure 22.
Visible VIs and browser access list.

file. Because reading an html file is not effected by version of web browsers. On the contrary, current version of Microsoft Office Word or Excel in your computer may not be compatible with LabVIEW **Report Generation** toolkit you installed.

The following VI generates an html report (**Figure 32**). Here, **Random Number** function generates Y array. X Array is generated by absolute time values.

When you execute the VI above the following report will be generated (**Figure 33**).

Report Generation toolkit also contains **Report** Express VI and **MS Office Report** Express VI (**Figure 34**). These Express VIs allow you to present data in form of html, MS Office Word or Excel.

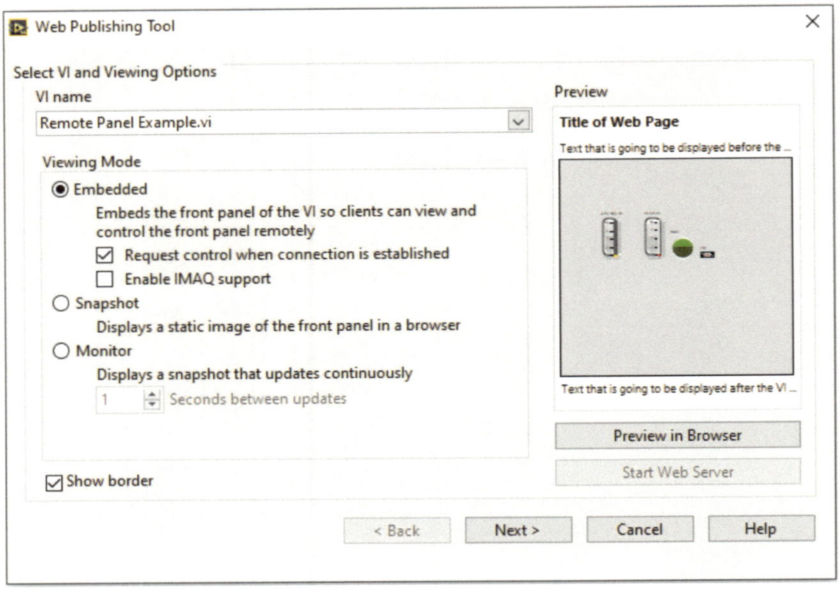

Figure 23.
Select VI and Viewing options.

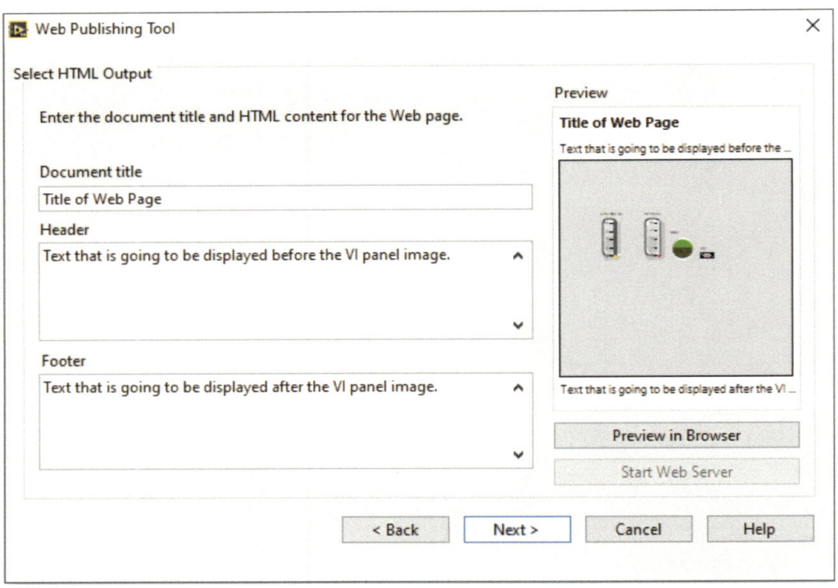

Figure 24.
Document title, header and footer.

In the following example, we use **Report** Express VI to present data in html format (**Figures 35** and **36**). You can also send data to printer or present data in MS Office Word or Excel format with the same VI. To do this double click **Report** Express VI and select the corresponding line from **Destination** tab in **Configuration Report** window.

When you execute the VI above you will see the following report.

DOI: http://dx.doi.org/10.5772/intechopen.96130

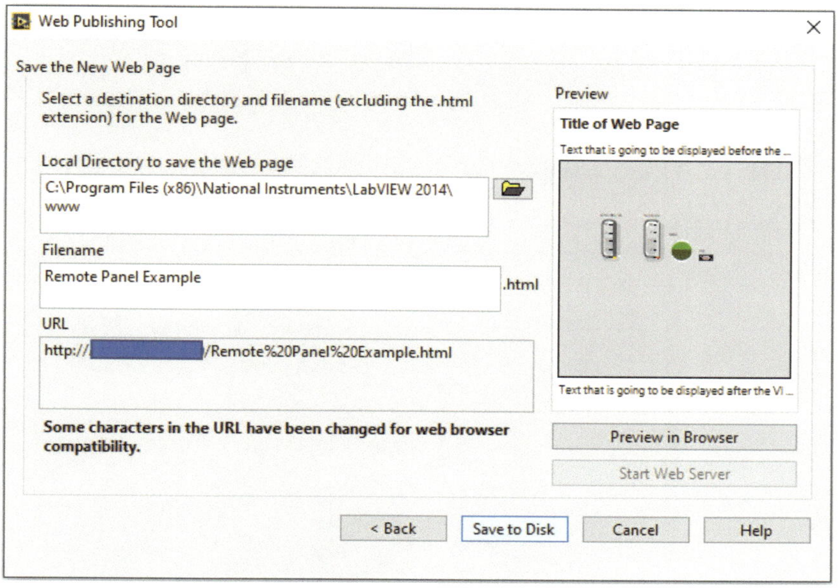

Figure 25.
Save the new web page.

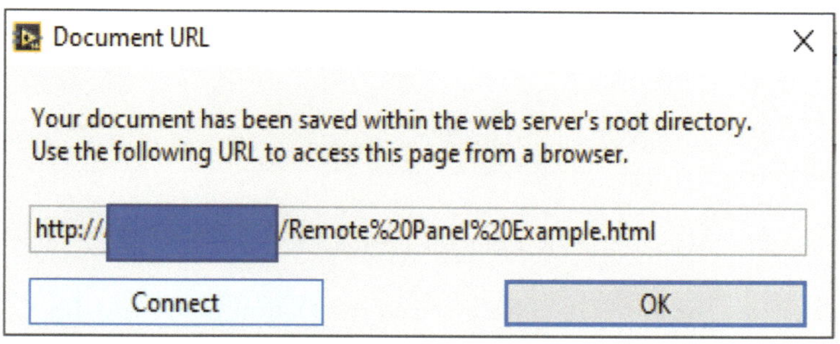

Figure 26.
Document URL window.

3.4 Save data

You can create folder, file or path, write and read data by using the **File I/O** VIs and functions (**Figure 37**). LabVIEW allows you to save data with different data formats.

Write Delimited Spreadsheet.vi converts a 2D or 1D array to a text string and writes the string to a new byte stream file or appends the string to an existing file. Both 2D and 1D arrays can be strings, signed integers, or double-precision numbers.

You should put **Write Delimited Spreadsheet.vi** out of the loop (**Figure 38**). Putting the VI inside the loop is not the good way of using it. Because LabVIEW at every iteration would open-write-close the file if it is inside the loop. This is not good in terms of VI efficiency.

You can also write date/time information for each data point. In the following example (**Figure 39**), data consist of time and random number.

Read Delimited Spreadsheet.vi reads a specified number of lines or rows from a numeric text file beginning at a specified character offset and converts the data to a 2D, double-precision array of numbers, strings, or integers. In the following

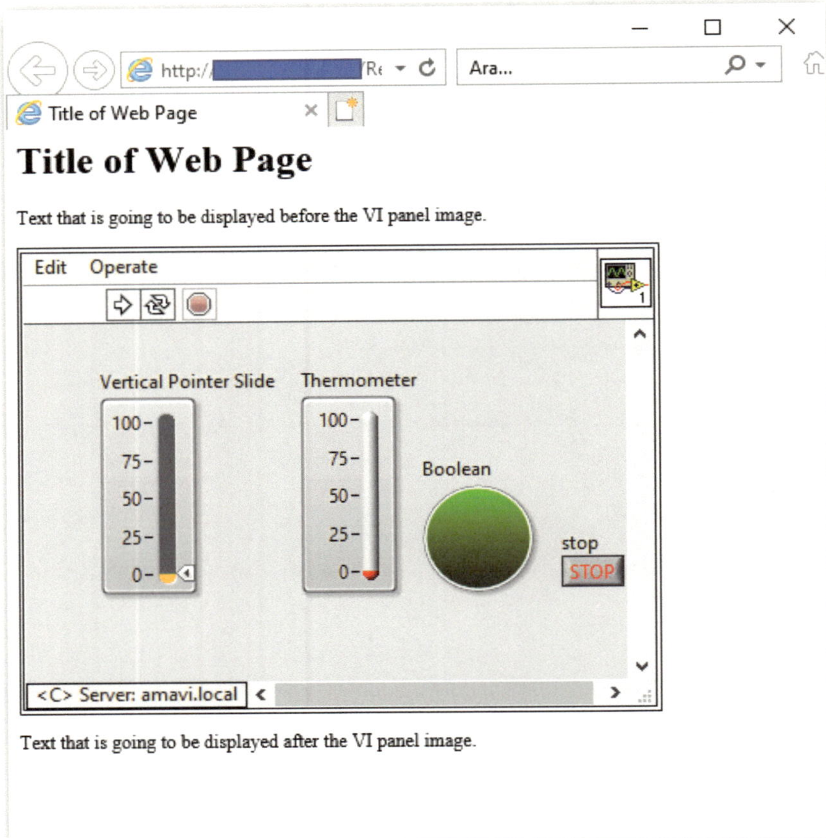

Figure 27.
Viewing and controlling front panels remotely with internet explorer.

example VI writes and reads data by using Write Delimited Spreadsheet.vi and Read Delimited Spreadsheet.vi (**Figure 40**). Note that we formatted time and random number by using **Format Into String**.

Another way to write and read data is to use **Write To Measurement File** and **Read From Measurement File** (**Figure 41**). It can only accept numeric or waveform data although **Write To Spreadsheet File** can accept array of strings, signed integers, or double-precision numbers. We recommend that you use **Write To Measurement File** to write data on disk. Because this Express VI allows you to save data as text (LVM), binary (TDMS), binary with XML header (TDM) and Microsoft Excel (.xlsx) formats.

Write To Measurement File is an Express VI. When you double-click to the Express VI **Configure Write To Measurement File** window opens (**Figure 42**). Here, you can configure writing.

As you see in **Figure 42 Configure Write To Measurement File** window allows you to change settings. It may be difficult to understand how you can configure this Express VI for the first time. To understand the function of each setting, we suggest that you individually experience with each setting.

In the following example we add time values (x) to the signal (random numbers, y) by using **Write To Measurement File** (**Figure 43**). Time data must be connected to **Comment** terminal of **Write To Measurement File.**

You can also save string data using **Write to Text File**. In the following example, VI saves date with time information (**Figure 44**).

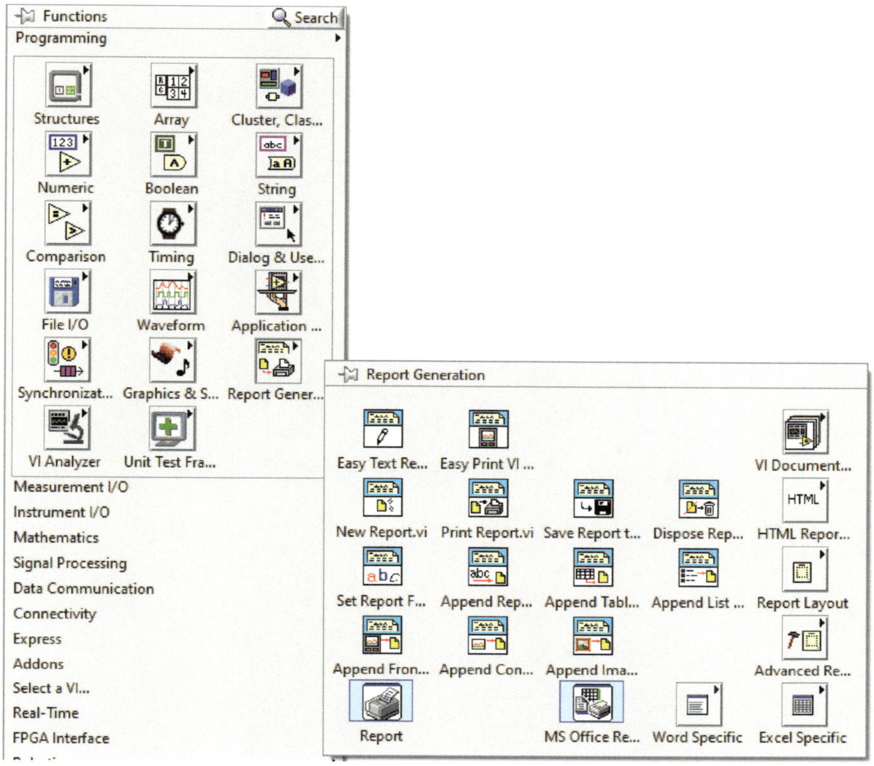

Figure 28.
Report generation palette.

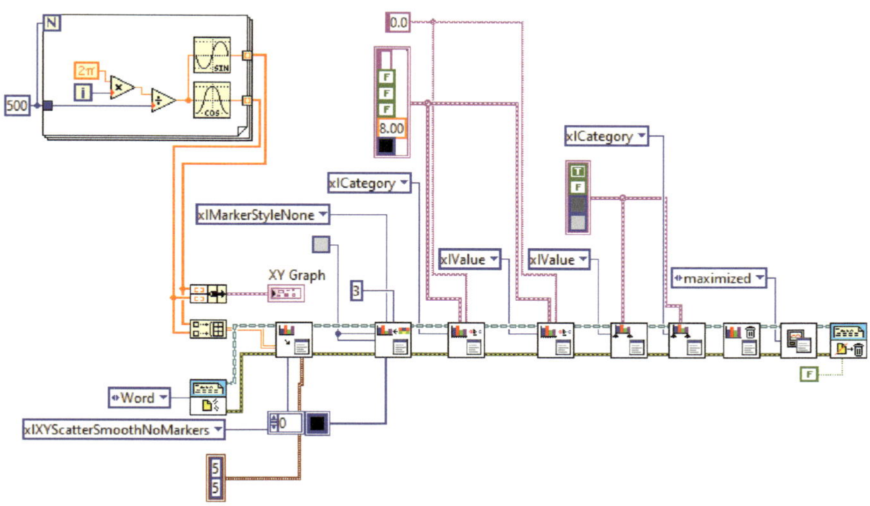

Figure 29.
Word report example VI.

Similar to examples above you can write and read data by using **Write Binary File** and **Read Binary File** (**Figure 45**). Binary files use less data storage. Therefore, it is useful when you have large data. However, **Write To Measurement File** save data as binary format, too. You may not need to use **Write Binary File** and **Read Binary File**.

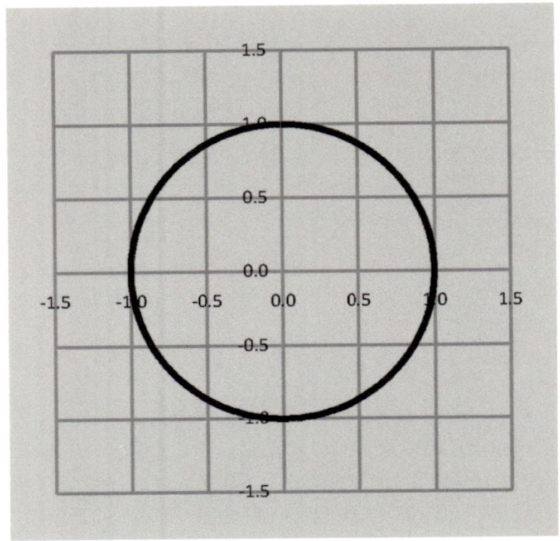

Figure 30.
Circle created by the VI.

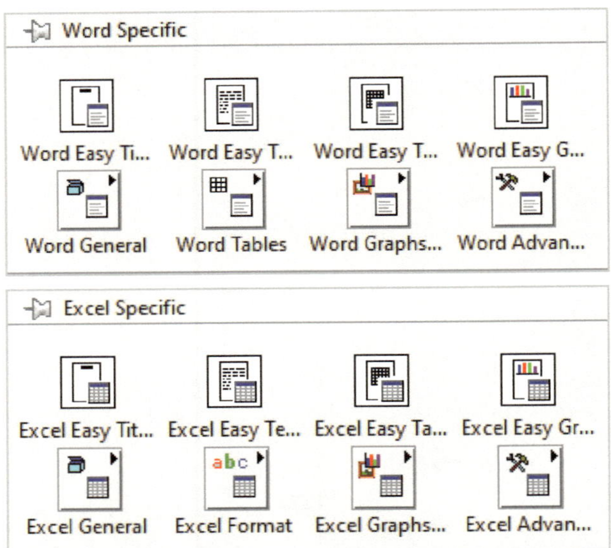

Figure 31.
Word specific and excel specific palettes located in report generation palette.

You can write and read waveform data by using the function of **Waveform File I/O** in **File I/O** (**Figure 46**).

The following VI generates two waveform sinus signals and write-read them (**Figure 47**).

NI has created a technical data management (TDM) solution. **TDM Streaming** is located in **File I/O** palette (**Figure 48**).

NI recommends costumers to use TDMS file format because it combines the advantages of several data storage options in one file format (**Table 1**). You can also work with TDM and TDMS files in Excel by utilizing the free TDM Excel Add-in for Microsoft Excel (supported Excel version: from 2007 to Excel 2016). You can take additional information from the following link [3].

DOI: http://dx.doi.org/10.5772/intechopen.96130

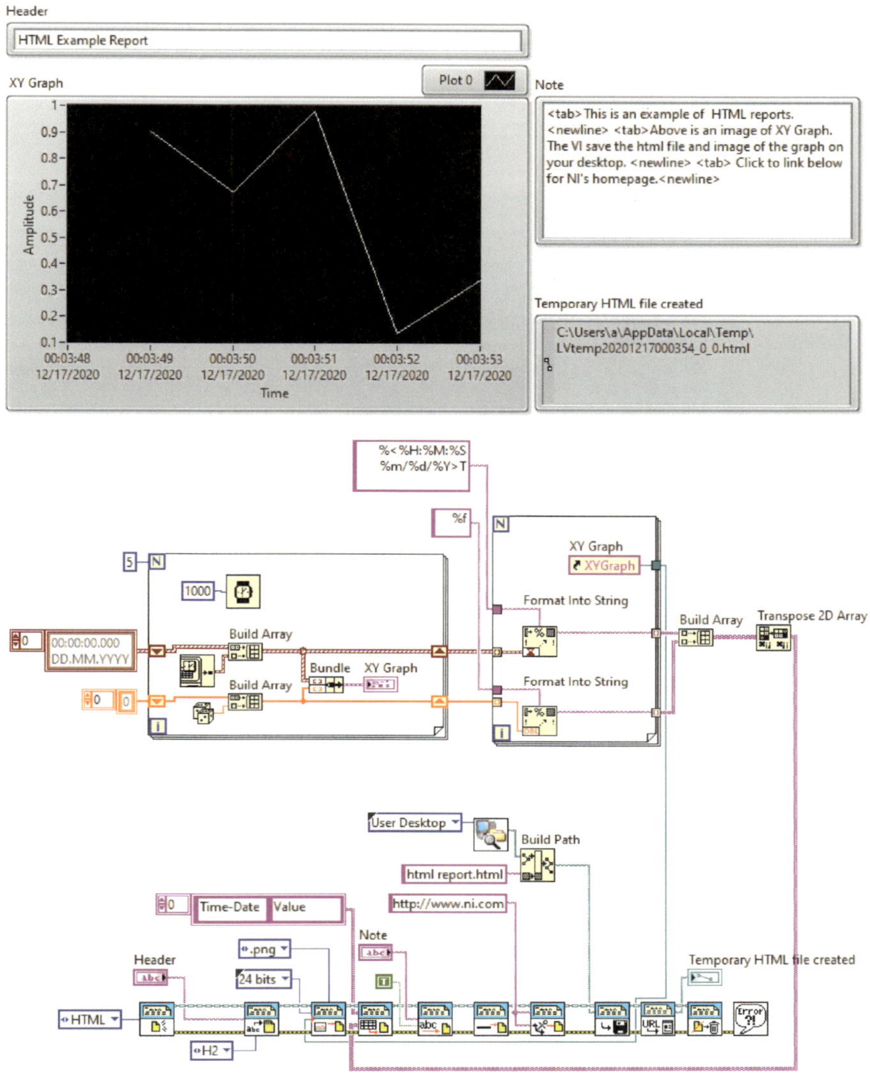

Figure 32.
VI generating an html report.

https://www.ni.com/en-tr/support/documentation/supplemental/06/the-ni-tdms-file-format.html.

3.5 Interactively manage data

3.5.1 National instruments DIAdem

DIAdem is a software to manage large amounts of data for measurement data aggregation, inspection, analysis, and reporting (**Figure 49**). With DIAdem, extraction of information from data can be efficiently performed. DIAdem is well adapted to LabVIEW. You can transfer your data from your LabVIEW application to DIAdem. With DIAdem, DataPlugins can be used to read, inspect and search different kinds of custom file formats. NI supplies free downloadable DataPlugins for hundreds of the most commonly used data file formats.

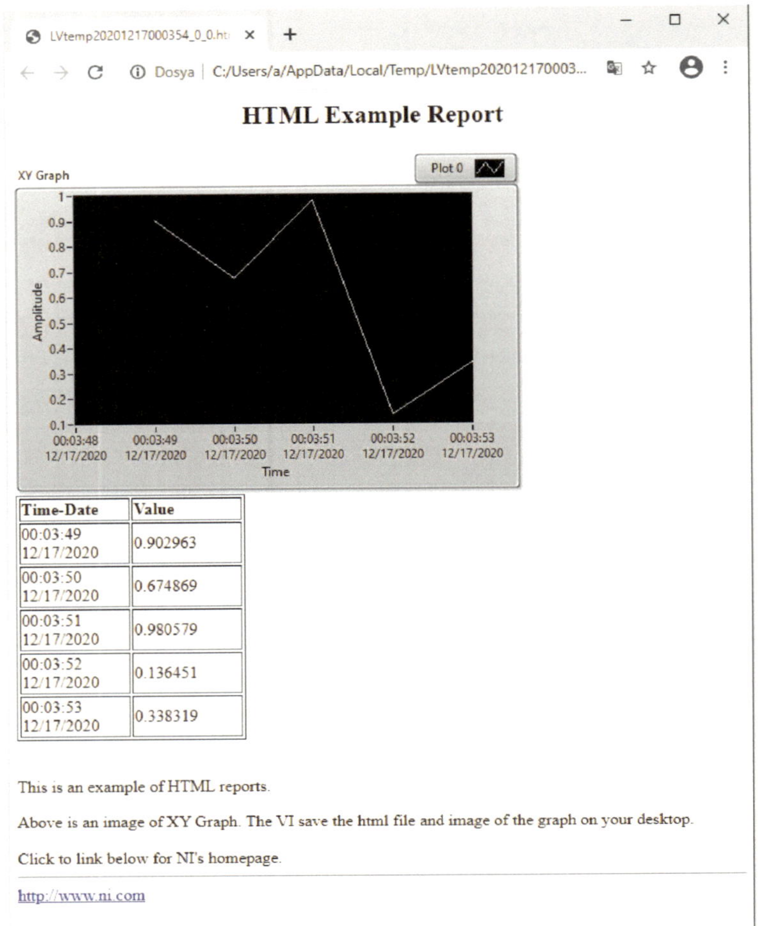

Time-Date	Value
00:03:49 12/17/2020	0.902963
00:03:50 12/17/2020	0.674869
00:03:51 12/17/2020	0.980579
00:03:52 12/17/2020	0.136451
00:03:53 12/17/2020	0.338319

Figure 33.
The html report generated by functions in report generation toolkit.

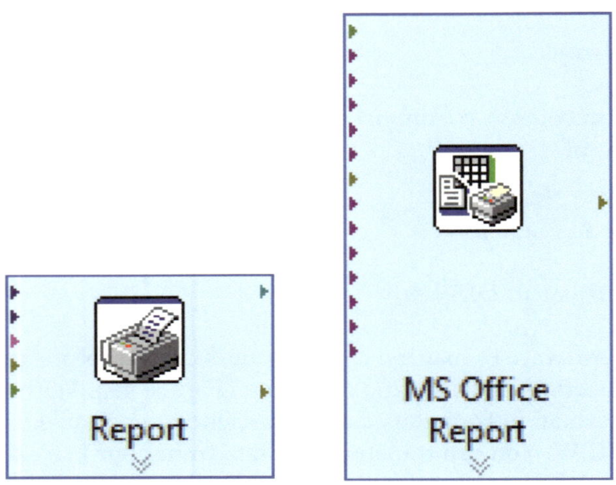

Figure 34.
Report express VI (left) and MS Office report express VI (right).

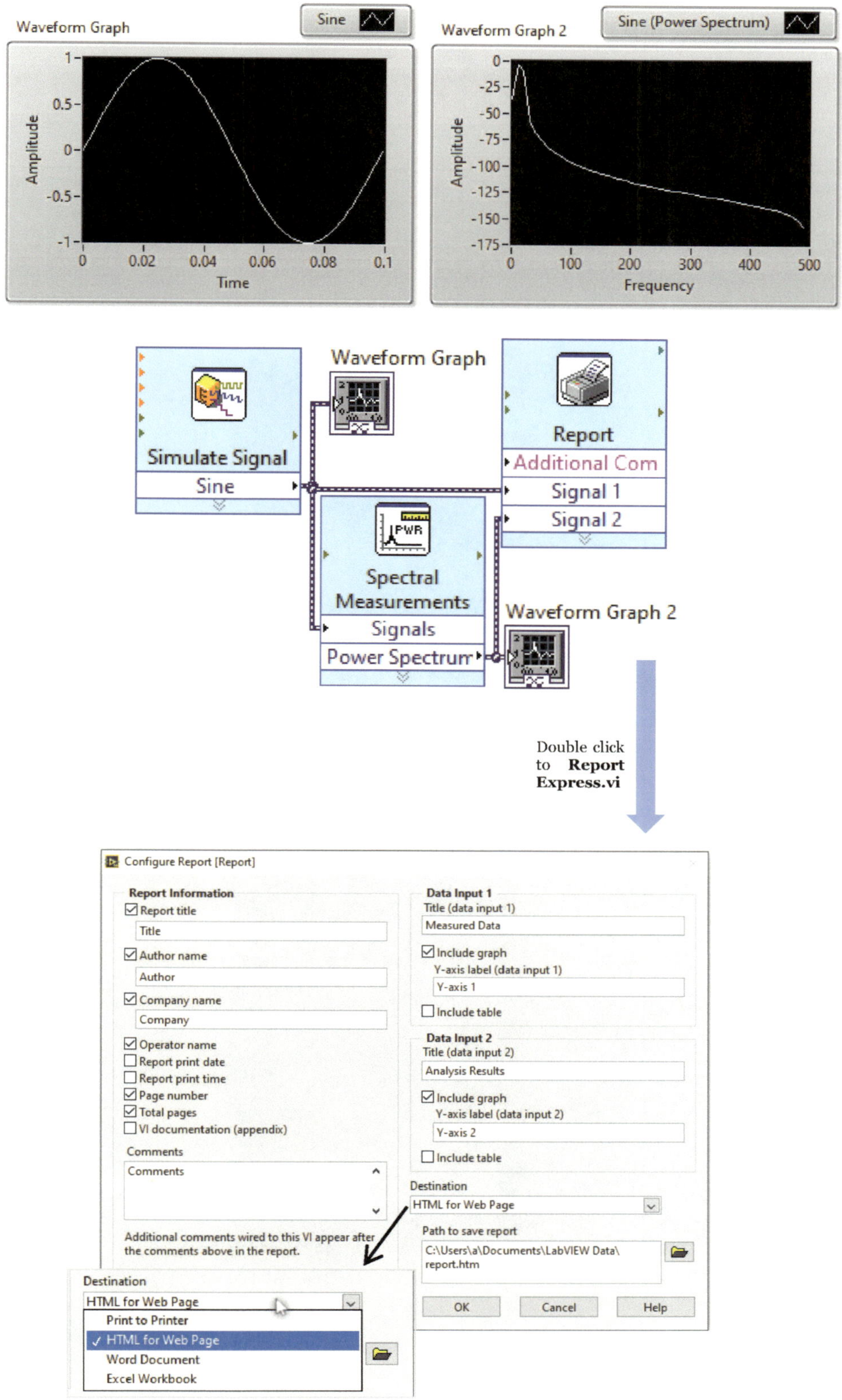

Figure 35.
Report express VI and Configuration report window.

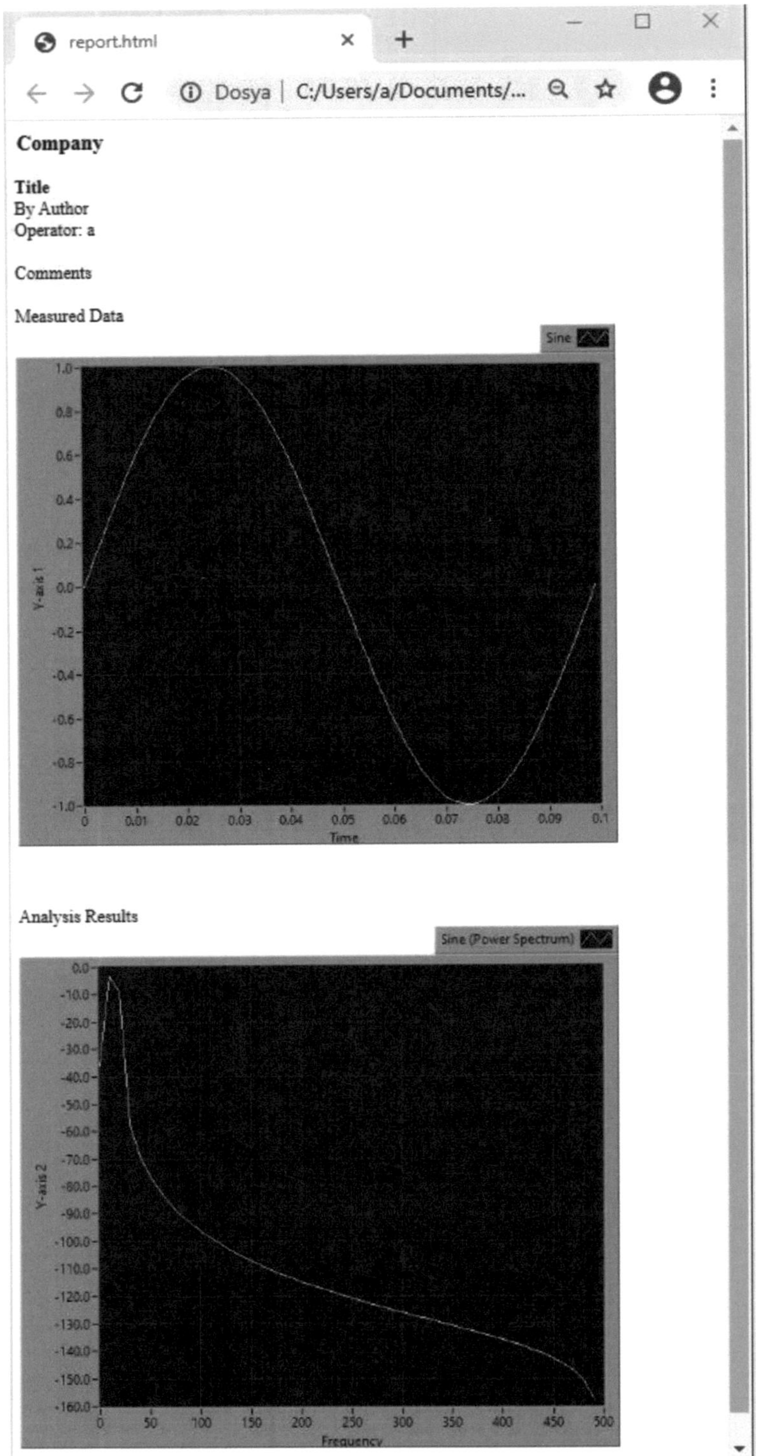

Figure 36.
The html report generated by report express VI.

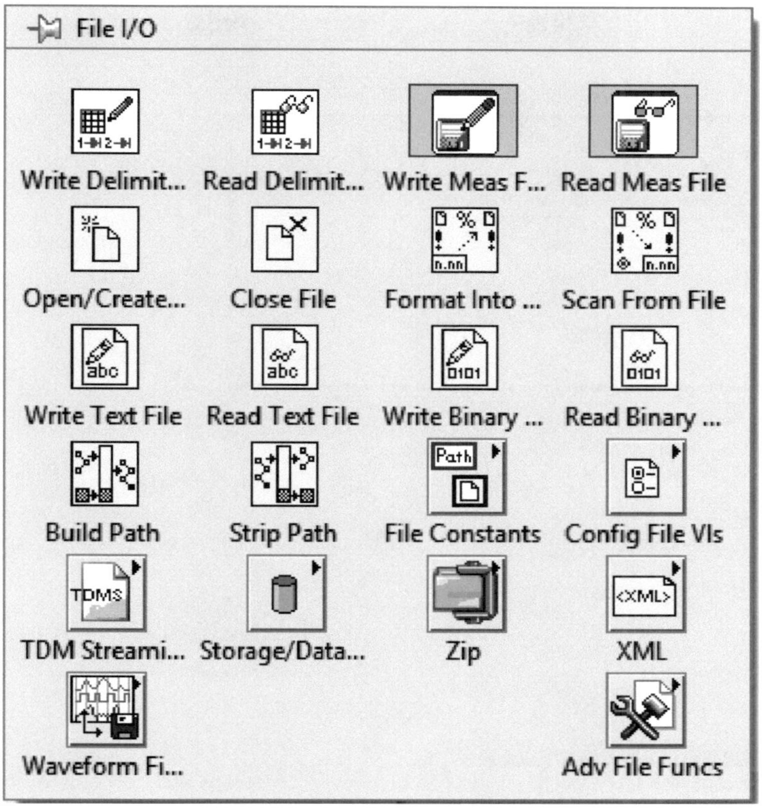

Figure 37.
File I/O palette.

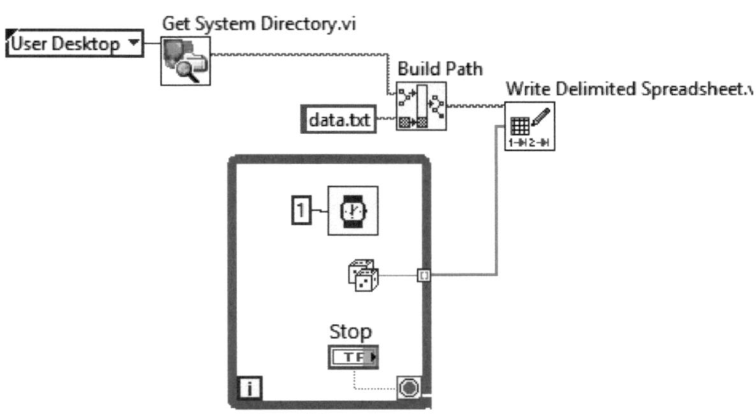

Figure 38.
VI can save data after loop is stopped.

3.5.2 LabVIEW and OriginPro

National Instruments engineers have created various NI LabVIEW add-ons which contain many functions and subVIs to meet the required functionality.

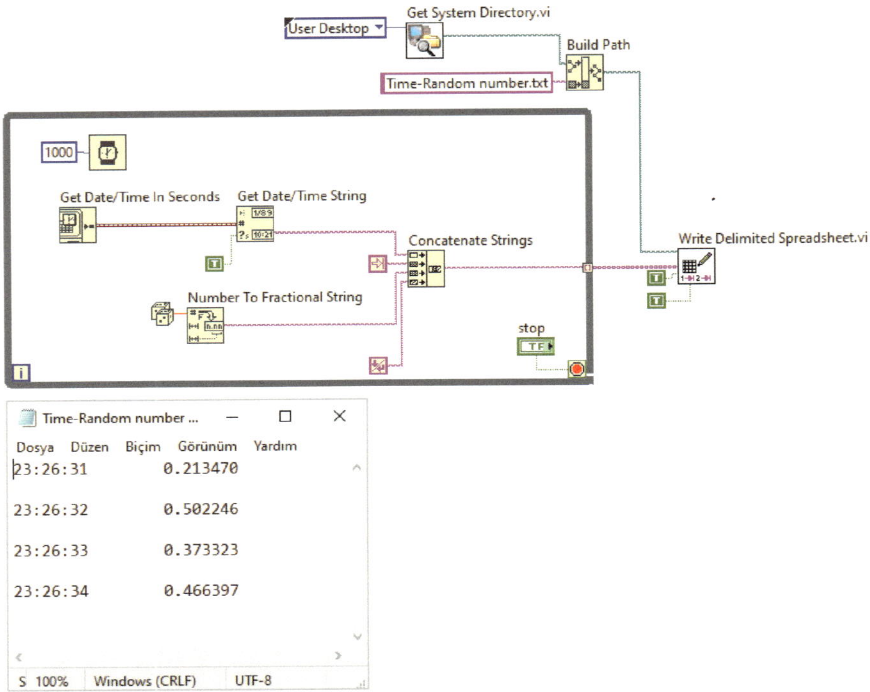

Figure 39.
Write delimited spreadsheet. Vi and resulted txt file.

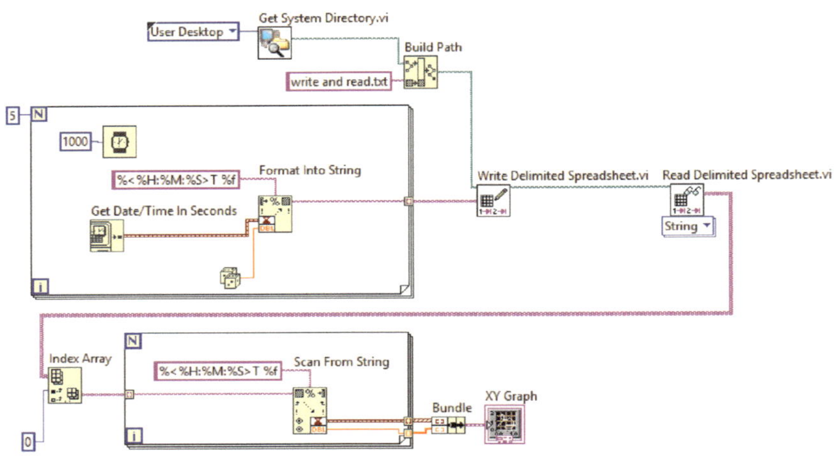

Figure 40.
Write and read data by using write delimited spreadsheet. Vi and Read delimited spreadsheet. Vi.

Besides the add-ons developed by NI engineers, add-ons have been developed for some other applications such as Origin. Origin or OriginPro is a powerful data analysis and graphics software preferred by scientists and engineers in industry, academia and research laboratories around the world [4]. Once the data collected by LabVIEW, the end-user will need to analyze the data and generate reports for presentation. Origin provides powerful analysis and graphing tools to reanalyze and present data. The ability to communicate easily between LabVIEW and Origin is a good platform that can greatly increase its efficiency in terms of data analysis and presentation [5]. There are studies in the literature using Origin and LabVIEW for

Figure 41.
Write to measurement file and read from measurement file.

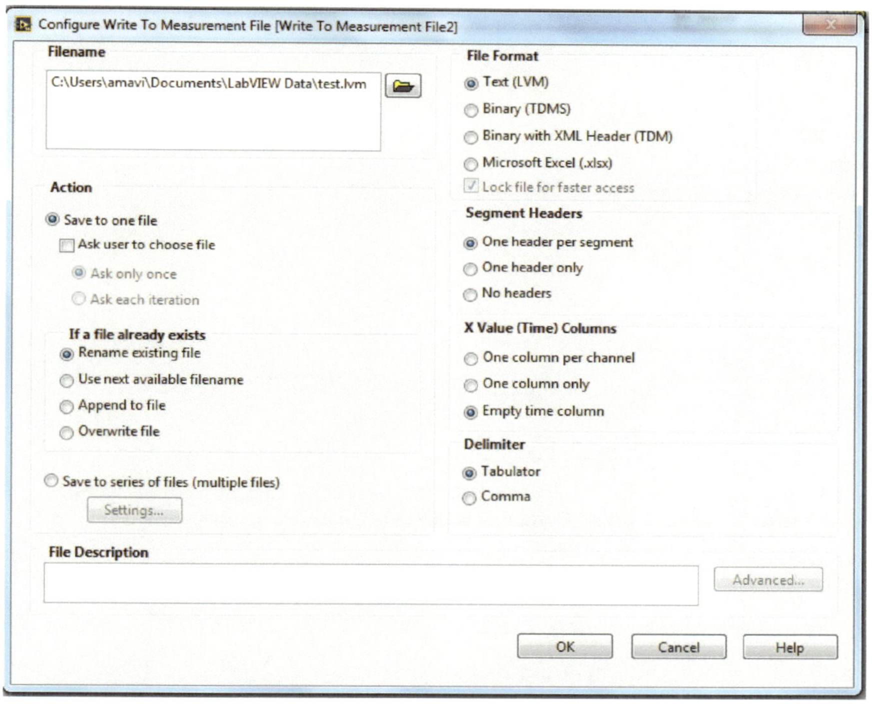

Figure 42.
Configure write to measurement file.

this purpose [6]. Origin provides subVIs to make work in LabVIEW environment. These subVIs allow to data transfer from LabVIEW to Origin and the data to be analyzed and presented in the Origin environment. This subVIs can be accessed from the folder where OriginPro is installed (Samples\COM Server and Client \LabVIEW). Also, in order to quickly access these subVIs in the LabVIEW environment, the subVIs can be copied from the installed folder and then pasted into a folder named OriginPro in vi.lip\addons folder where LabVIEW is located [7]. OriginPro library can be accessed from the menu opened by right-clicking on add-ons in the Block Diagram window in LabVIEW as shown in **Figure 50**.

After adding the subVIs provided by OriginPro to LabVIEW, you can see subVIs as shown in **Figure 51**. There are four sections under Origin function palette. These are

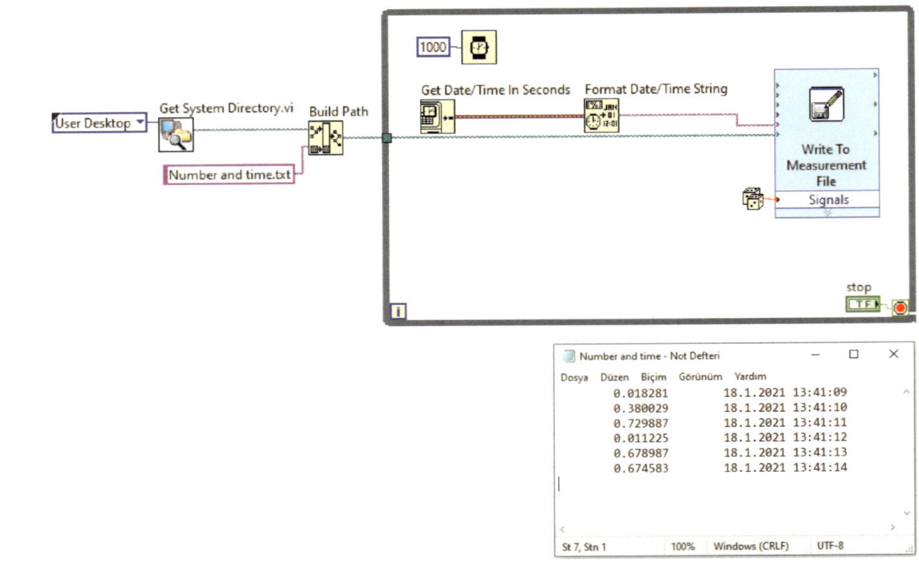

Figure 43.
Write to measurement file with time information.

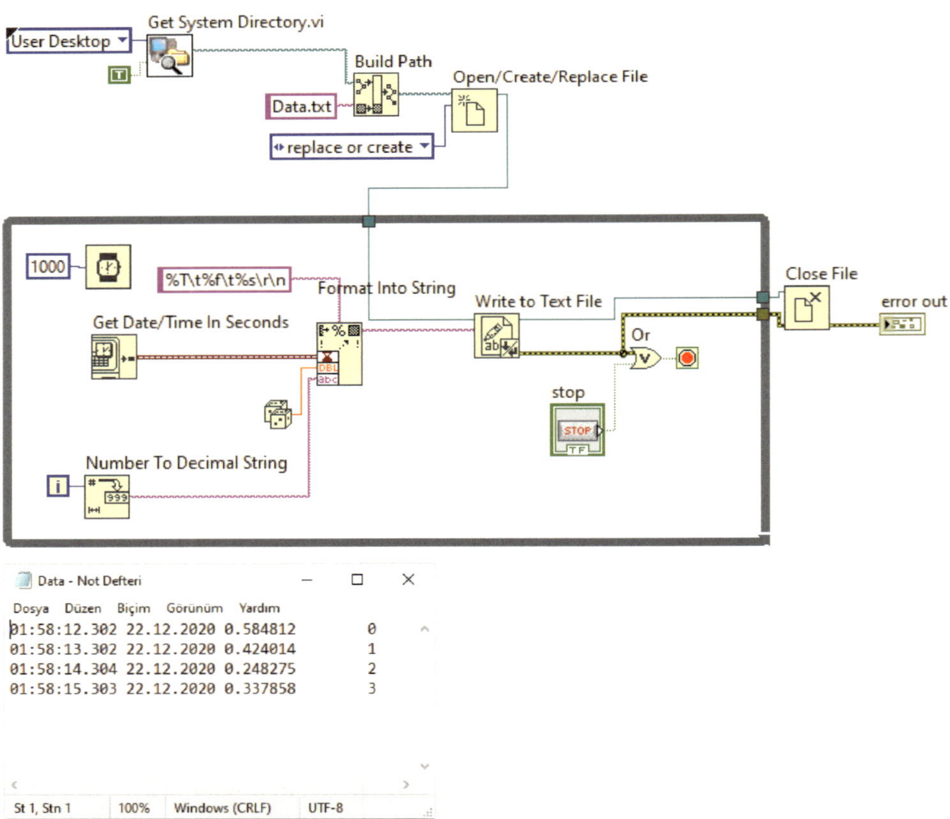

Figure 44.
Write to text file and resulted txt file.

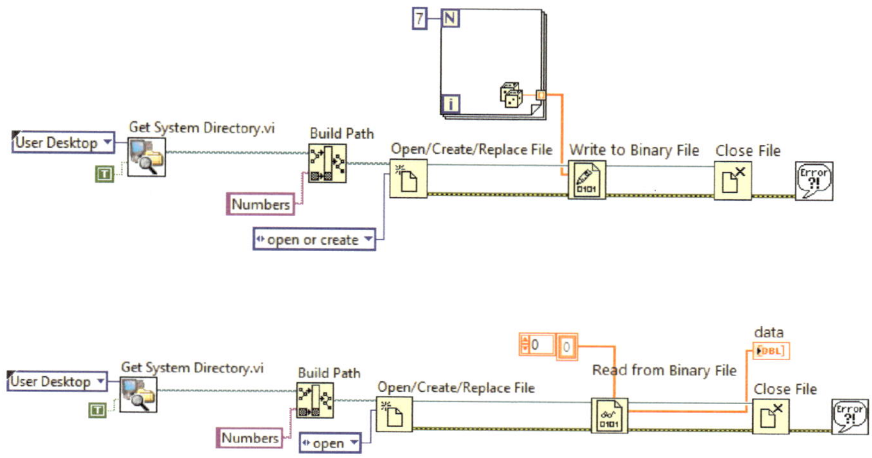

Figure 45.
Write to binary file and read from binary file.

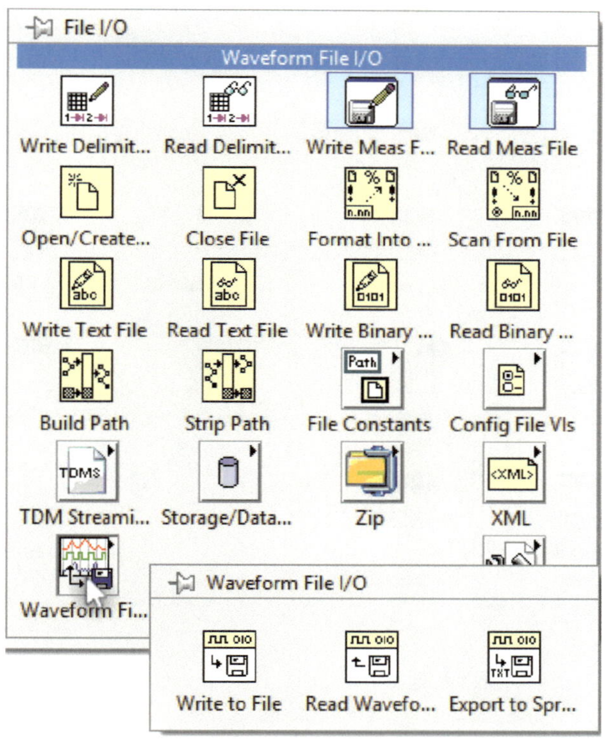

Figure 46.
Waveform file I/O.

- OriginApp: Basic VIs that handles the Origin OPJ files, worksheet and columns,
- OriginAppClassics: Older VIs existed before Origin 8 (deprecated),
- OriginWave: VIs that handles Origin matrix objects,
- OriginMatrix: VIs that handles LabVIEW Waveform data.

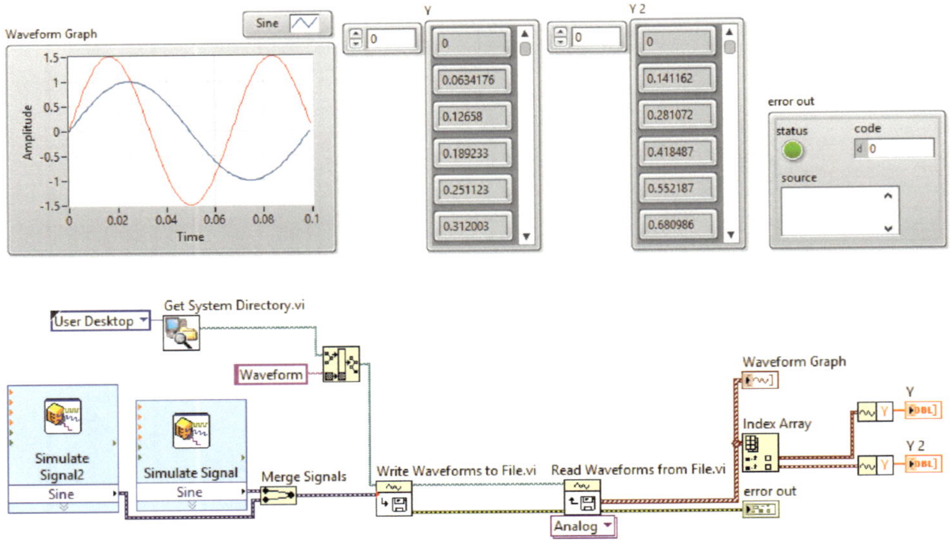

Figure 47.
Write-read waveforms.

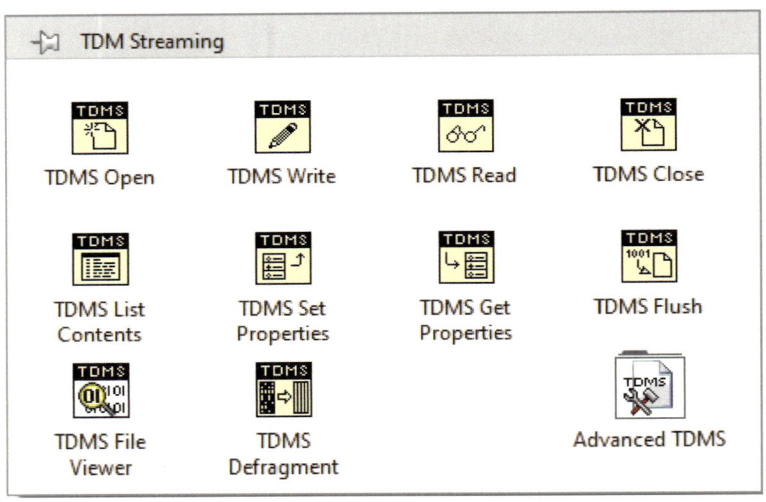

Figure 48.
TDM streaming.

	ASCII	Binary	XML	Database	TDMS
Exchangeable	√		√		√
Small Disk Footprint		√			√
Searchable				√	√
Inherent Attributes			√		√
High-Speed Streaming		√			√
NI Platform Supported	√	√	√	√	√

Table 1.
Advantages of TDMS file format.

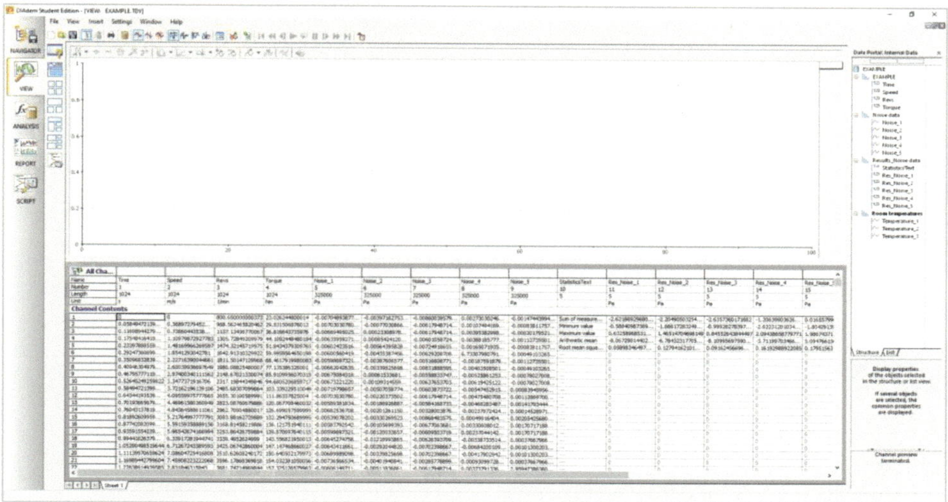

Figure 49.
DIAdem.

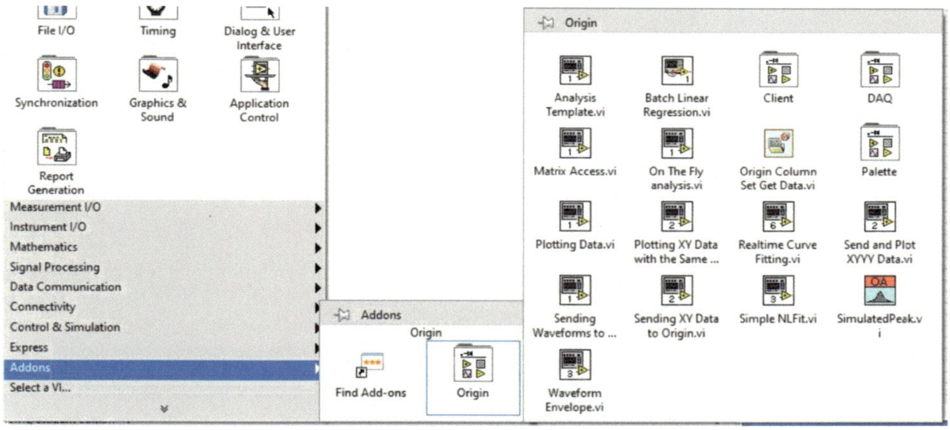

Figure 50.
LabVIEW library of OriginPro.

In the following example, the measured temperature values are transferred and plotted in OriginPro. In this example, first, add **OA_ConnectToOrigin.vi** to the Block diagram so that LabVIEW can connect to OriginPro. Once the link between LabVIEW and OriginPro is established, add **OA_NewWorksheet.vi** to the Block diagram to open a new worksheet in OriginPro. This also creates the name and details of the worksheet to use the entries of VI. After that, you use **OA_GetColumn.vi** to select a column. Then, you send the information of this column via **OA_Col-Setting.vi**. You will have to pay attention to two important points while filling in the entries of this VI. The first is the Data Format. This is the part where you need to write the format of the data. The second is the column type. Here we can determine the axis of the column to use in the chart. You can specify an axis such as X, Y, or Z, as in the example shown in **Figure 52**. After this process is completed, you will be able to transfer your data to the worksheet by using **OA_Col-SetData.vi**. You can use the **Read Delimited Spreadsheet.vi** in the File I/O menu to send the data to worksheet created in the Origin environment as in the example shown in **Figure 52**. You should pay attention for the correct format of the data. In addition, if you are working with data consisting of a single data created in

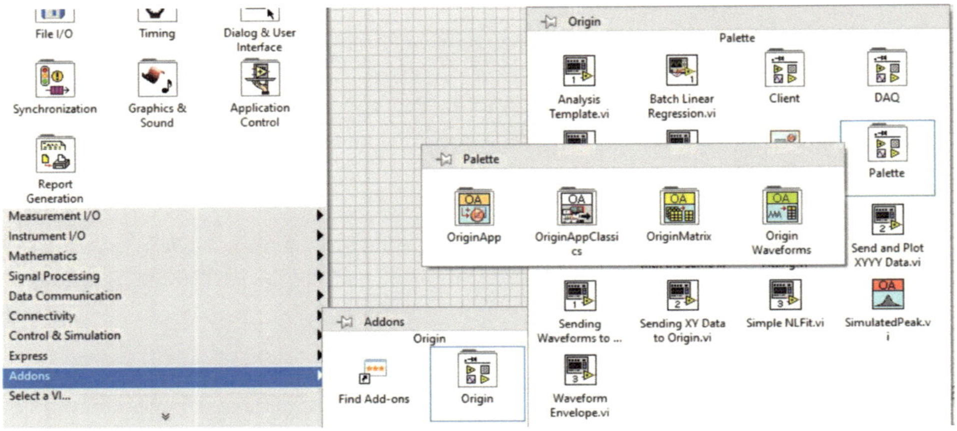

Figure 51.
Palette section.

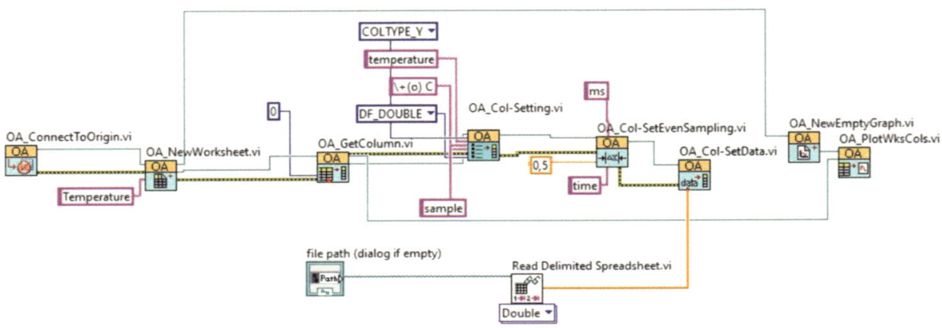

Figure 52.
Sending data from LabVIEW to OriginPro.

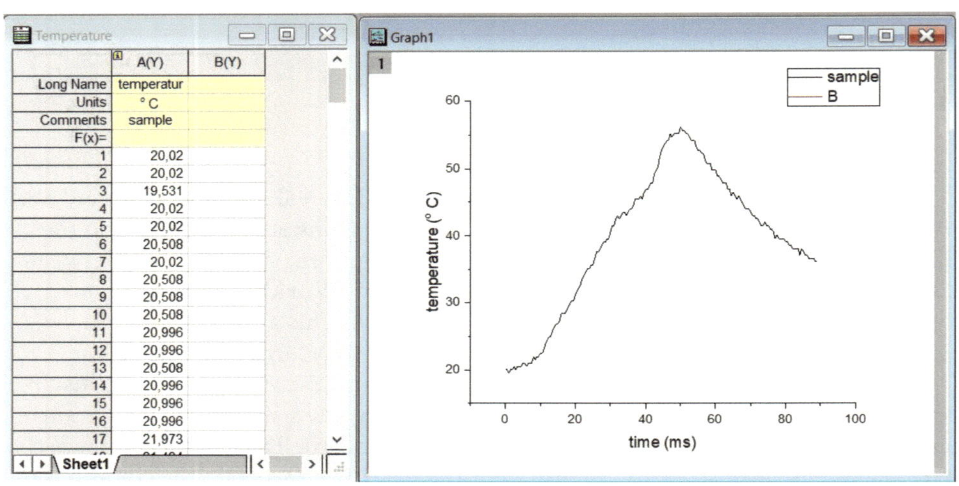

Figure 53.
The output of the OriginPro.

certain periods of time such as the data used in this example, and if you are going to plot these data versus time, you will need to use **OA_Col-SetEvenSampling.vi**. Finally, VI in **Figure 52** plots the temperature data at 0.5 ms intervals as shown in **Figure 53** by using **OA_NEWEmptGraf.vi** and **OA_PlotWksCols.vi**.

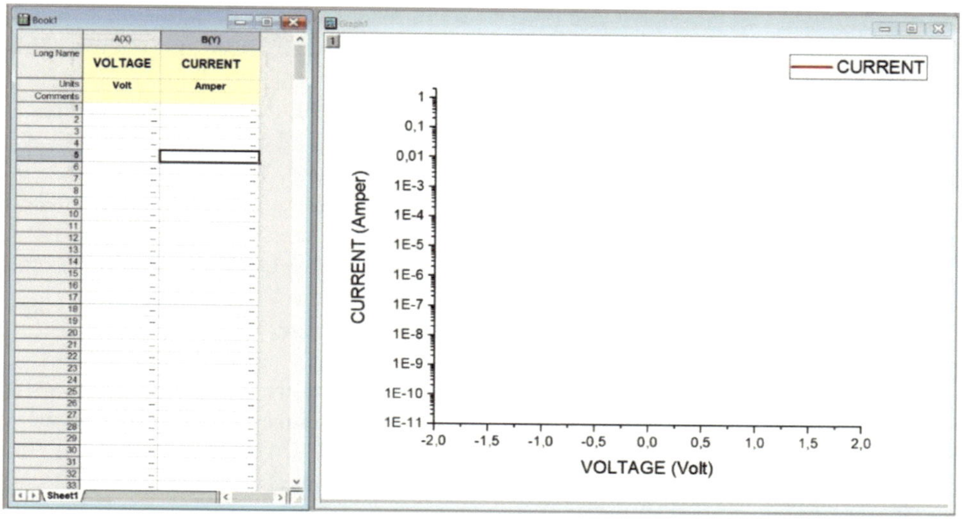

Figure 54.
Origin template.

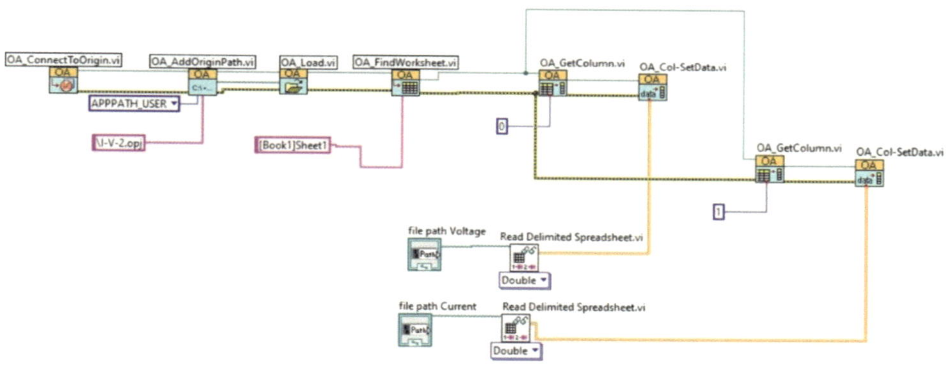

Figure 55.
Utilization of origin templates in LabVIEW.

Similarly, another example is to use the template you created in the OriginPro. For this, you must first create a template in the OriginPro. Save this template in Documents\OriginLab\2015\User Files folder in your computer. In this case, it will be easy for you to access this template from LabVIEW. For an example application, a template has been prepared in OriginPro as in **Figure 54**. Now add **OA_ConnectToOrigin.vi** to the Block Diagram for the connection between Labview and OriginPro. **OA_AddOriginPath.vi** creates the file path. Then, use **OA_Load.vi** to upload file. **OA_FindWorksheet.vi** is to select worksheet. After this step is completed, since you defined Voltage for the X axis and Current for the Y axis in the template, you need to send your data to OriginPro accordingly. For this, you can use **OA_GetColumn.vi** to select column. You can send your data to these columns with the help of **OA_Col-SetData.vi**. These steps are given in **Figure 55**. Here, the data you previously saved in the LabVIEW environment is opened again and drawn in OriginPro [8].

4. Conclusion

The scientist and engineers frequently need to measure physical changes, analyze them, and present data from the measurement. LabVIEW includes hundreds of built-in and add-ons functions for analysis and presentation of data. If you do not know LabVIEW well, you should check the examples in LabVIEW and ni.com. This is the easiest way to learn how you can analyze and present your data.

LabVIEW has effective 2D (Charts and Graphs) and 3D visualization tools for data presentation. Graphs accept only array data. It plots all the received points at once. Charts attach received data to already existing points. When an array of points is wired to a chart or graph, LabVIEW assumes the points are equally spaced out. If you also want to define X axis values, you should use XY Graph.

LabVIEW allows you to present your data html, Microsoft Office Word or Excel. LabVIEW also allows you to publish any application to the Web with Remote Panels. Therefore, your VIs can be reachable as a Web page. You can control a remote device from your home. It is interesting, right?

Generally, you want to save and read your data. There are different ways to do it in LabVIEW. We recommend you to use **Write To Measurement File** and **Read From Measurement File**. These supports data as text (LVM), binary (TDMS), binary with XML header (TDM) and Microsoft Excel (.xlsx) formats.

DIAdem software is a NI product to manage data for measurement data aggregation, inspection, analysis, and reporting. Interestingly, DIAdem can use more than one thousand data file formats by utilizing DataPlugins.

More than 500,000 clients in the world use OriginPro to import, graph, explore, analyze interpret their data. If you are a user of OriginPro software, you can integrate it with LabVIEW. When you install Origin add-ons for LabVIEW you can easily communicate with OriginPro.

Thanks

We thank to Serdar Bölükbaşıoğlu, manager of **Ludre Software** company, for his contribution.

Author details

Ahmet Mavi[1], Ahmet Özmen[2] and Mehmet Ertuğrul[3]*

1 Department of Nanoscience and Nanoengineering, Atatürk University, Institute of Science, Erzurum, Turkey

2 Department of Electronics and Automation, Vocational School, Agri Ibrahim Cecen University, Ağrı, Turkey

3 Department of Electric and Electronic Engineering, Faculty of Engineering, Atatürk University, Erzurum, Turkey

*Address all correspondence to: ertugrul@atauni.edu.tr

References

[1] Analysis Concepts [Internet]. 2004. Available from: https://www.ni.com/pdf/manuals/370192c.pdf [Accessed: 2020-12-27]

[2] Mavi A, Ertuğrul M. LabVIEW Uygulamaları 1 (in Turkish). 1th ed. Ertual Akademi; 2016 308 p. ISBN:978–605–83722-4-5

[3] The NI TDMS File Format [Internet]. 2020. Available from: https://www.ni.com/en-tr/support/documentation/supplemental/06/the-ni-tdms-file-format.html.html [Accessed: 2020-12-27]

[4] Origin and OriginPro [Internet]. 2021. Available from: https://www.originlab.com/Origin [Accessed: 2021-01-18]

[5] Use of National Instrument LabVIEW as a client application [Internet]. 2021. Available from: https://www.originlab.com/doc/Origin-Help / LabVIEW-client-app [Accessed: 2021-01-18]

[6] Weiskirchen R, Weiskirchen S, Kim P and Winkler R. Software solutions for evaluation and visualization of laser ablation inductively coupled plasma mass spectrometry imaging (LA-ICP-MSI) data: a short overview. Journal of Cheminformatics. 2019;11:16. DOI: 10.1186/s13321-019-0338-7

[7] Using Origin from LabVIEW [Internet]. 2016. Available from: www.originlab.com/pdfs/Origin2017_Documentation/English/Using_Origin_from_LabVIEW_E.pdf [Accessed: 2021-01-18]

[8] Özmen A, Coşkun A, Ertuğrul M. Measurement and Analysis with LabVIEW Software. In: Academic Studies in Engineering. HAYALOĞLU ADNAN; GÜNDAY ABDURRAHMAN; 2020. p. 95–106.

Chapter 2

TeraVision: A LabVIEW Software for THz Hyper-Raman Spectroscopy

Rohit Kumar, Qiucheng Yu, Domenico Paparo and Andrea Rubano

Abstract

Terahertz Time-Domain Spectroscopy (TDS) has emerged during the last two decades as a very popular technique for characterizing the low-energy excitations of several materials, gaseous, liquids and solids, as well as artificial materials as for instance epitaxial heterostructures and more. In recent years, the advances in THz technology allowed obtaining nonlinear optical effects with THz photons, showing remarkable results. In particular, THz Hyper-Raman Spectroscopy greatly expands the spectroscopic capability of the standard THz-TDS by combining intense and broad-band THz pulses with a detailed analysis of the spectral content of the generated signal. It is evident that this improvement needs an adequate software support. The main parameter for coding the software which differs with respect to a standard THz-TDS software is the control of a motorized grating (monochromator), but several routines employed in the setup optimization stage rather than the actual measurement are needed as well. In this paper we present the TeraVision software, based on LabVIEW code, in order to highlight the solutions we adopted to tackle the main experimental challenges as well as to give a pleasant and user-friendly experience to expert users.

Keywords: nonlinear spectroscopy , THz time, domain spectroscopy, THz hyper, Raman spectroscopy , optics, pump&probe

1. Introduction

1.1 Brief introduction on terahertz radiation

Terahertz (THz) is a multiple of the unit of frequency in the International System, equal to 10^{12} Hz. In the present context we use it for representing the frequency of electromagnetic waves in the range of 0.1 to 30 THz (wavelength 10 μm to 3000 μm), which confines with microwaves and infrared (IR) light. The THz radiation lies on a transition region from macroscopic classical theory to microscopic quantum theory and from classical electronics to photonics. Sometimes, especially in the past, people refer to this region with the expression "Terahertz gap". The reason for this is that both the generation and detection of THz electromagnetic waves is quite challenging compared to well-established techniques in the microwaves and infrared regions, so that this important part of the electromagnetic spectrum was inaccessible until very recent years. Since the first artificial THz wave was produced and detected by means of ultrashort laser pulses in the early 90's, a strong research effort began with the aim to improve our ability to produce, control, manipulate and detect this kind of waves [1], so that now, 30 years later, the expression "THz gap" is not so often employed anymore.

The importance of THz spectroscopy in fundamental science is obvious because many important low-energy excitations are present within this range in practically all kind of materials: emerging materials [2], semiconductors [3], ferroelectrics [4], superconductors [5], polymers [6], photonic crystals [7], liquids [8–10], gases [11] and biological systems [12]. Moreover, THz technology has a high impact in many different areas too. We shall not cite them all here, but as an example the application of THz sensing and imaging to biomedical and biological studies and laboratory practices will revolutionize, in our opinion, these fields in a close future [13]. One of the great advantages in this case, is the very low photon energy, which is much safer than X-rays for living tissues imaging [14].

The unique performance of THz has far-reaching implications in areas such as communications [15], radar [16], astronomy [17], safety inspection (airports) [18], etc. Terahertz is a new source of radiation with many unique advantages: as an example, it is possible to combine spatial and temporal resolution (a typical THz pulse has a time duration of about 1 ps and a wavelength of about 100 μm) in order to perform not only imaging but also time-resolved measurements on an ultrafast ps time-scale. As a result, THz research has a great value to national economy and national security [19].

1.2 THz Spectroscopy

For many years, IR and Raman spectroscopies have been used to investigate the properties of molecular vibrations and rotations. THz Spectroscopy can complement those two techniques, for mainly two reasons: from the one hand, it can access higher wavelengths compared to IR and Raman, and from the other hand it can access the complex transmission or reflection coefficients rather than their real squared value (power spectrum), as will be shown in the following [20].

Figure 1 shows the position of the THz range in the electromagnetic spectrum. At present, there is no generally accepted definition of the upper and lower limits of THz radiation frequency. The high frequency of THz spectrum overlaps with the far IR spectrum, and the low frequency overlaps with the microwave frequency band. Most of the authors set the limits of this range to 0.1–3 THz, because this is the emission region of the most commonly used sources (photoconductive antennas and optical rectification), but here we prefer to extend this range up to 30 THz, because this is the upper limit of our source based on air-plasma, which will be described in the following, and this emission scheme is essential to achieve the THz Hyper-Raman effect, as it will be shown later.

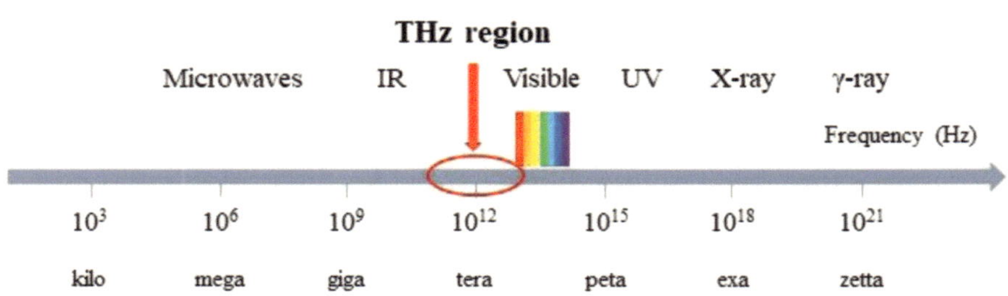

Figure 1.
The terahertz region in the electro-magnetic spectrum.

A major reason for the rapid development of terahertz spectroscopy is the progress of ultrafast lasers (femtosecond pulses) by means of the Chirped Pulse Amplification technique, which was introduced by Donna Strickland and Gérard Mourou at the University of Rochester in the mid-80's for which they received the Nobel Prize in Physics in 2018 [21]. These ultrashort laser sources gave birth to several new generations of spectrometers in the early 90's, including THz. The THz spectrometer is capable of generating and detecting pulses of coherent THz radiation, but most importantly it is able to detect both amplitude and phase of the THz wave, and thus it allows retrieving the full complex dielectric function of the target material without resorting to Kramers-Kronig relationships, as it happens in standard IR spectroscopy. Moreover, some recent advances in both the fs laser sources and the THz emission efficiency and detection sensitivity, allowed the researchers to make use of intense THz pulses to control the material relevant parameters [22], rather than being limited to measure its spectrum.

1.3 THz time domain spectroscopy

THz Time Domain Spectroscopy (THz-TDS) is the most widespread technology among general THz spectroscopies. We give here a brief description of it, although it is not the main topic of this Chapter, for two different reasons. First, the TeraVision software is able to control standard THz-TDS measurements besides the THYR Spectroscopy, and, second, THz-TDS is a good way for inexperienced readers to approach the topic of THz Spectroscopy in general. The name of this technique is quite straightforward as the spectroscopic information is not retrieved directly in the frequency domain, but in the time domain, and then converted by simple Fourier Transform. The THz pulse can be generated in several ways. The most commonly employed generation mechanisms are photoconductive antennas [23], optical rectification [24] and air-plasma four waves mixing [25]. The first method creates a strong and short burst of current in between two electrodes with applied bias voltage when a population of free charges is suddenly created by a fs-laser optical pulse in a semiconductor slab. The second method converts the fs-pulse bandwidth into THz by nonlinear optical difference frequency conversion ($\omega-\omega$, often called "optical rectification", this expression more properly refers to static electric fields generated by light) by means of opportune nonlinear crystals (ZnTe, GaSe, GaP and many others have been proved to work best on different spectral ranges). The key point here is that due to the very large bandwidth of fs-pulses the difference between all possible frequencies ($\omega-\omega$) does not always vanish, but it rather creates a low-energy band whose extension depends on the laser pulse bandwidth as well as the dielectric properties of the nonlinear crystal. The third method is based on laser ionization of air molecules, converted by strong laser intensity into plasma, which is a highly non-linear medium. The contextual presence of fundamental and double frequency photons (created by a suitable nonlinear crystal Beta Barium Borate β-BBO, placed few centimeters in front of the plasma) generates many four-waves mixing frequencies: some of them are falling into the visible domain, as possible to observe in **Figure 2** on the mirror surface, and should be filtered out by a high resistivity thick silicon wafer, but one particular linear combination of those frequencies ($2\omega-\omega-\omega$) is capable of producing the THz radiation. Also, here as in previous point, the subtraction of two photons at fundamental laser frequency from one photon at double frequency, is not a static field ($\omega=0$) because of the large bandwidth of the fs-laser pulses. This is the generation scheme which is necessary for THYR Spectroscopy [26].

For detection, it is possible to use any of the above mentioned techniques, but reversed. In all cases, the detection works as follows: the THz pulse is sent to the sample and it gets reflected or transmitted. Then, it is sent to a detector which will

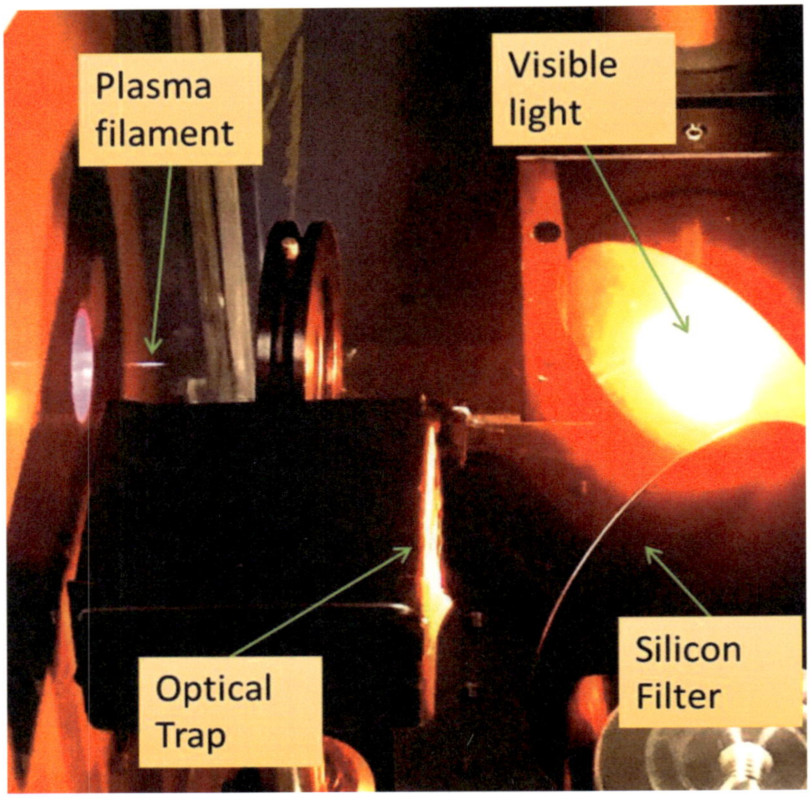

Figure 2.
Plasma filament for THz generation.

change its electric or optical behavior according to the THz excitation, and namely to the value (and sign) of the THz electric field at a specific time. As an example, the THz pulse will change the microcurrent which flows in between two electrodes on a semiconductor substrate (photoconductive antennas) but this current will exist only when a second fs-laser optical pulse hits the semiconductor to create free carriers. Given the much shorter duration of the fs pulse as compared to the THz one, the microcurrent will be proportional not to the integrated THz pulse energy, but rather to the value of the THz electric field in time, where the time is given by the delay between the THz and the probing optical pulse. In a similar fashion, one can exploit the transient birefringence of certain nonlinear crystals in order to change the polarization state of the probing optical pulse. Again this transient change depends on the time delay between THz and optical pulses and therefore it gives a time-scan of the temporal profile of the THz pulse (so-called "electro-optic sampling"). Finally, the symmetry breaking induced by the THz electric field in a centro-symmetric medium (usually air) will produce a nonlinear effect when coupled with a strong optical pulse, such as for instance optical Second Harmonic Generation (SHG), and again its intensity will be proportional to the amplitude of the THz electric field responsible for the symmetry breaking (this is the so-called Air Biased Coherent Detection - ABCD).

THz-TDS has the following characteristics:

1. THz-TDS technology can be used for qualitative identification [27], because it can effectively detect the physical and chemical information of materials in the THz band. And it is also a non-destructive detection method.

2. The amplitude and phase information of various materials such as dielectrics [28], semiconductors [3], gas molecules [11], biological systems [12, 29] (proteins, DNA, etc.) and superconductors [5, 30] can be obtained conveniently and quickly by using THz-TDS technology.

3. In conductive materials, THz radiation can directly access the carrier dynamics. THz-TDS can effectively be used as a non-contact and ultrafast multimeter for conductivity measurements.

4. Due to the transient nature of THz radiation, THz-TDS techniques can be used to measure relaxation-times and the low-energy excitations dynamics in general, with very good resolution [31].

Besides, THz-TDS technology also has the advantages of wide bandwidth, high detection sensitivity and stable operation at room temperature.

Despite a huge amount of developments in the last 20 years, THz-TDS still needs some improvements on the spectral resolution and spectral range. In the near future, THz-TDS technology will be a powerful tool for uncovering and analyzing ultrafast phenomena in basic sciences such as physics, chemistry, and biology. At the same time, with the reduction of laser cost, the emergence of more efficient THz emitters and detectors, and more compact and advanced optical design, THz-TDS technology will have a broad commercial application prospect [32].

1.4 THz Hyper-Raman time domain spectroscopy

The standard THz-TDS technique has already proved to be very effective. Nonetheless, it is important to highlight that, despite the use of nonlinear optics to generate and reveal THz radiation, this is a linear spectroscopic technique, i.e. it probes the linear dielectric function of the target material. Sometimes, it could be not so easy to access the target observables, such as the low-frequency excitations (phonons, magnons, excitons, etc.). For example, a large absorption can obscure a significant part of the spectral range, the index of refraction and/or the dielectric function could be difficult to interpret if the target materials have a complicated structure, the background from the bulk can overcome the weaker superficial contribution, which is perhaps the real target of the experiment, and finally some low-energy excitations are simply out-of-reach because of symmetry, i.e. they just do not show up in the linear parameters due to selection rules. Accessing the nonlinear optical parameters can in principle remove most of these hurdles. With this aim in mind, a new spectroscopic technique has been developed by our group a couple of years ago. Through the application of this new technology, femtosecond laser pulses and intense sub-picosecond broadband THz pulses produce THz-optical four-wave mixing in the material. The spectrum of the resulting signal is decomposed both in wavelengths and time and it appears as two distinct frequency side bands around the optical SHG central frequency $2\omega_L$, where ω_L is the optical central frequency of the fundamental fs-pulse. This effect resembles the well-known Hyper-Raman (HYR) effect, and therefore it has been named THz Hyper-Raman - THYR.

In standard all-visible HYR, the coupling between optical frequencies happens through a high-order nonlinear susceptibility tensor. The model at the base of the THYR effect, instead, is a regular four-wave mixing effect, in which one of the optical photons is replaced by a THz photon. The mathematical formalism is therefore that of a nonlinear Raman effect, but because of the very small THz photon energy, the sidebands are appearing very close to the SHG central frequency

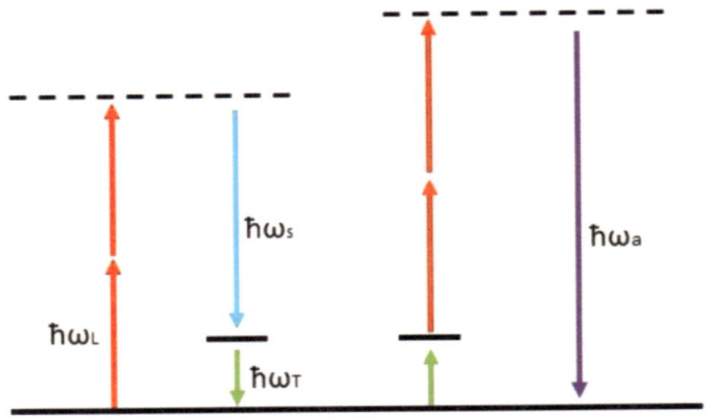

Figure 3.
Level diagram of THYR effect. Labels are: 1) L = laser 2) T = terahertz 3) S = stokes 4) a = anti-stokes.

instead of the fundamental central frequency, resembling what happens in HYR. Taking the same nomenclature routinely used for Raman and HYR effects, we define "Stokes" and "anti-Stokes" bands, developing around the SHG frequency $2\omega_L : \omega_{s;a} = 2\omega_L \mp \omega_T$ (where ω_T is the THz central frequency. For all details about THYR theory and applications we invite the reader to the following publications: [33, 34]). The energy diagram of the effect is shown for Stokes (left) and anti-Stokes (right) bands in **Figure 3**. From an experimental point of view, the signal has to be measured in both frequency and time domain, so that the software TeraVision which has been developed in order to drive the experiment and record the data must be more sophisticated than the regular kind of software used for THz-TDS. Moreover, we intended to create a software capable of performing several different kinds of measurements, thought to be independent from the THYR spectrometry, and/or used as ancillary measurements in THYR spectrometry as well. In addition to this, we inserted some features in order to simplify the signal detection and optimization, which could be of use more generally for whoever needs to perform any kind of optical Pump/Probe experimental scheme.

2. Experimental set-up

The experimental setup which was built in order to perform THYR experiments is shown in **Figure 4**. The femtosecond laser is a Ti:Sa mode-locked seed laser, amplified by a regenerative cavity. The fs-laser pulse, after passing through the beam splitter, is divided into a Pump (transmitted) pulse and a Probe pulse (about 10% of total power). The Pump is chopped by a mechanical chopper locked to half of the laser trigger frequency in order to block every second pulse. The pulse train chopping is very important because it allows measuring the difference between THz-ON and THz-OFF signals, as it will be further explained in the following. The pulse is then sent on a nonlinear optical crystal (Beta Barium Borate, β-BBO) where about 20% of the optical power is converted in SHG and then both fundamental and doubled pulses are focused in air by an achromatic doublet. In the plasma filament, usually about 1 cm long, the generated plasma is producing THz pulses via four-waves mixing processes in which the maximum THz amplitude depends on relative phase between the fundamental and second harmonic light [35]. The Probe is sent to a delay stage in order to introduce a controllable delay between Pump and Probe pulses and then it is incident on the detection crystal (usually ZnTe, GaSe or GaP, or just air for ABCD technique) together with the THz pulse in a collinear geometry, through a hole in the last parabolic mirror. In the detection crystal the polarization

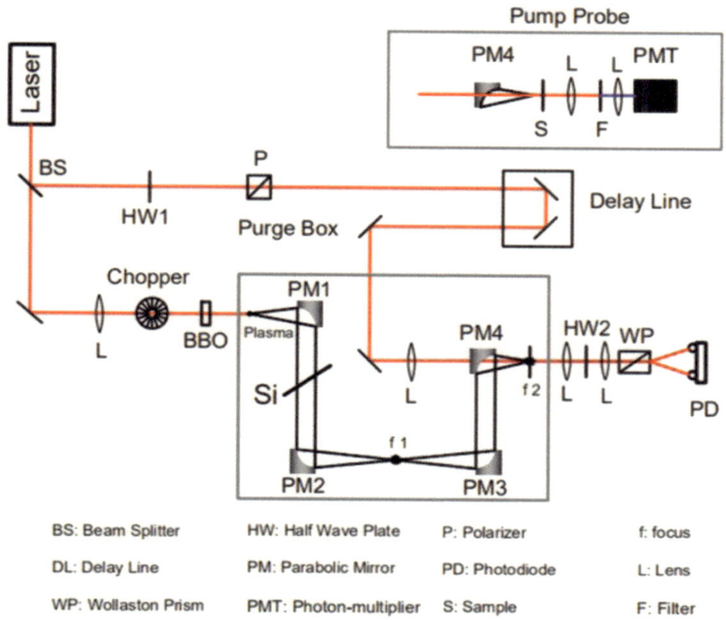

Figure 4.
Experimental setup of THz-TDS and THYR.

state of the optical pulse is altered by the presence of a transient birefringence created by the THz electric field, and this change can be detected by separating two orthogonal polarizations of the pulse with a Wollaston Prism and measuring their signal difference in a balanced photodiode. As the delay between the two pulses is adjusted, the amplitude of the THz pulse can be sampled in time, and the full THz waveform can be reconstructed. In a standard THz-TDS experiment, the sample will be placed in the first focus (f1) of the parabolic mirror and the detection crystal is placed in the second focus (f2), as we can see from **Figure 1**, to measure the THz waveform transmitted through the sample. The complex transmission coefficient is then obtained by measuring the signal without the sample in place, as an estimate of the incoming electromagnetic wave, and finally the complex ratio between the Fourier Transforms of those two measurements will deliver the complex transmission coefficient spectrum. In reflection geometry, the problem is more complicated by the difficulty to place sample and reference mirror with sufficient accuracy. We are not going to tackle the problem of THz reflectivity measurements here, and we refer to this publication and the references therein [36]. In the THYR experiment, the first focus (f1) is left empty, the detection crystal is removed, the sample is place in the second focus (f2) and the whole detection line after sample is replaced (by means of flippable mirrors) with the setup shown in the inset of **Figure 4**. The signal produced in the sample by the interaction between THz and optical pulses is filtered in order to remove the 800 nm light, then it is sent to a monochromator for spectral analysis and finally to a photon multiplier tube (PMT) for detection. The PMT is needed because of the very small amount of photons which are created by THYR effect, but its operation is usually in continuous mode rather than in photon counting mode. Anyway, if the signal from a given sample would be too small, it is straightforward to switch the detection to a photon counting regime, provided that the laser and setup stability is good enough to support a much longer data acquisition time. Finally, the ON and OFF trains of pulses are separated by the TeraVision software and each of them is subtracted with the following one. As the time-distance of two subsequent pulses is 1 ms, this differential measurement scheme will quench all noises below the 1 kHz cutoff frequency. Moreover, this procedure ensures that even in presence of a strong non-THz-related background signal (as for instance a static SHG signal, or 2-photons luminescence) the measured signal will not be too much affected.

Here and in the following, we will name the Pump pulse simply "THz pulse", as it is responsible for THz generation, and the Probe pulse will be named "Gate pulse", as it sets the time-gate at which the signal is sampled. This nomenclature is quite common in THz spectroscopy, and it is particularly important when performing a THz Pump/Probe measurement, because in that case the pulses will be three: an optical Pump pulse which is exciting the sample, the THz and the Gate pulses, which are both together acting as a Probe. Let us now briefly give some details about the main components of the setup.

2.1 Laser

The laser system used to run the spectrometer is a Coherent Legend regenerative amplifier seeded by a Coherent Mantis fs-oscillator (800 nm central wavelength, 20 fs FWHM pulse duration, 80 MHz repetition rate, 500 mW output power) and pumped by Coherent Evolution (527 nm central wavelength, ~ 10 ns pulse width, 1 kHz repetition rate, 20 W output power). The Legend delivers ~4 W output power at 1 KHz repetition rate (~ 4 mJ energy per pulse) at 800 nm central wavelength (1.5 eV), ~ 80 nm bandwidth, and with about 35 fs FWHM time duration [37].

2.2 Signal detection

2.2.1 Monochromator

It is an optical device which disperses the light spectrum by means of a suitable optical grating (1800 groves/mm, in this case) and filters out all wavelengths except those passing through a slit at the output of the device. We use a monochromator from the Optometrics Group which is a Ebert-Fastie type with focal length (path length) of 74 mm, model number SDMC1–02 (Scanning Digital MiniChrom). The wavelength range of this monochromator is 200-800 nm. The wavelength range is suitable for all our measurements (THYR, THz-TDS and SHG). The reciprocal linear dispersion factor is 3 nm/mm. We can operate our monochromator manually using a knob for selection and a digital counter for wavelength readout. However, the device is routinely controlled via TeraVision software which drives a stepping motor controller (PCM-01) connected to PC by RS-232 serial port. The monochromator is used only for SHG and THYR detection, in combination with a PMT. It is possible to use it also for THz-TDS detection in case of ABCD technique, but we will not discuss this method here.

2.2.2 Photomultiplier tube and balanced photodiode

A PMT is a device consisting of a photocathode, several dynodes and one anode, all placed in vacuum. Modern PMTs deliver low noise and low jitter detection over a wide dynamic range. We use a PMT by Hamamatsu (model number R928) having spectral response from 185 to 900 nm. The PMT is used only for SHG and THYR detection, and it can be used in combination with the monochromator for THz-TDS by means of the ABCD technique. In case of standard electro-optic sampling (THz-TDS) we use a balanced amplified photodetector by Thorlabs (model number PDB210A) with broadband detection range (320-1060 nm). The detector has two distinct low-noise photodiodes, whose difference is strongly amplified by the device in order to achieve a low noise measurement of the differential signal. The main advantage of this device compared to performing the subtraction of independent photodiodes signals is that the amplification on the difference is affected by a much lower noise than the subtraction of already amplified separate signals.

2.3 Delay line

In our experiments we use a linear stage by Physik Instrumente (PI, M-5x1 series) which are low profile, high-accuracy linear translation stages for laboratory applications. The stage is controlled by a DC-Motor controller (Mercury) and the PC communication is ensured through a standard RS-232 port. The minimal step size corresponds to a minimum incremental motion of 0.1 μm and 1 μm full travel accuracy. Therefore, since the optical path is double (forward-and-backward) than the physical path, we can achieve a minimal time delay of about 6 fs. As the optical pulse duration is about 35 fs, this Delay Line allow us to reach the full time-resolution possible for our setup.

2.4 Rotors

To control the input and output optical polarizations of light for THYR and THz-TDS experiments, we used mechanical rotors controlled by DC servo controllers. We use Thorlabs rotors (model PRM1/MZ8) which are small, compact, DC servo motorized 360° rotation stages. The user can measure the angular displacement manually by using the Vernier dial and the marks that are marked on the rotating plate in 1° increments. In order to rotate the rotors precisely we are using DC servo controllers (model TDCOO1) or "T-cube DC", controlled in turn by our TeraVision software. The software allows the user both to set the in-out angles as fixed parameters in any kind of measurement, or to measure the signal as a function of in-and/or-out angles by fixing all other parameters, in the "Rotors" Tab.

2.5 Data acquisition

The electrical signal is generated by the PMT or the balanced photodiodes and enters a Stanford Gated Integrator or Boxcar Averager (GI). Here, it is splitted in two signals (50%). One is sent to a Tektronix DPO 4054 oscilloscope for monitoring the signal and, more importantly, adjusting the synchronization with the electronic gate, while the second signal enters the GI for processing. In our lab we use SR250 GI that is a versatile, high speed, low cost module designed to recover fast input analog signals from noisy background. It consists of a gate generator, a fast gated integrator, and an exponential averaging circuitry. The gate generator can be triggered internally or externally, and it provides an adjustable delay from a few nanoseconds to 100 ms before it generates a continuously adjustable gate with a width between 2 ns and 15 μs. The delay can be set by a front panel potentiometer. The fast gated integrator integrates the input signal during the gating and the output signal is normalized by the gate width to provide a voltage which is proportional to the average of the input signal. This signal can be further amplified by using different settings on the front panel of the boxcar integrator. The final analog output signal is sent to a data acquisition system (DAQ) based on aPCIe-6351 interface by National Instruments

(NI). The latter is a X-series multifunction I/O Device, which offers analog (16 bit) and digital I/O, four 32-bit counter/timers for encoder, frequency, event counting etc. It relies on a high-speed PCI Express bus. In our TeraVision software we used NI-VISA which is an API (Application Programming Interface) that provides a programming interface to control GPIB, serial, USB, PXI, VXI etc. instruments in National Instruments application development environments like LabVIEW.

3. The TeraVision software

3.1 Brief introduction to LabVIEW

LabVIEW (Laboratory Virtual Instrument Engineering Workbench) is a software development environment by National Instruments (NI). One of the most significant difference between LabVIEW and many other computer languages is the use of a graphical interface for editing codes, resulting in the form of block diagrams. Besides this, LabVIEW is quite straightforward to use especially for people (like us) with no professional training for software coding, thanks to many routines and libraries which have been developed with a special focus on laboratory work, automation and instruments control, together with a pleasant and user friendly graphic interface for the software end user. This software is the core of the NI design platform and it is also an ideal choice for developing measurement or control systems. LabVIEW development environment integrates all the tools that engineers and scientists need to quickly construct various applications, designed to help engineers and scientists to concentrate on their principal tasks, to solve problems and to improve productivity.

LabVIEW is a graphical programming environment for creating applications using icons instead of text lines. The traditional text programming languages determine the order of program execution according to the order of statements and instructions, while the LabVIEW adopts the method of data flow programming. LabVIEW's graphical source code is similar to a flowchart to some extent, so it is also called the block diagram. The data flow between nodes in the program block diagram determines the execution order of subroutines (Virtual Instruments – VI) and functions. LabVIEW provides many controls that look similar to traditional instruments, such as oscilloscopes and multimeters, for simplified user interface creation. The user interface in LabVIEW is called the front panel. With icons and wires in the block diagram, one can programmatically control the objects on the front panel.

3.2 Main concepts about the TeraVision software

The TeraVision software is meant to perform two main types of measurements: 1) as a function of time (THz-TDS); 2) as a function of time and wavelength (THYR). In addition, it can perform two ancillary measurements: 3) as a function of wavelength only (Spectrum); 4) as a function of input/output polarizations (Signal Anisotropy). The latter two can be performed with and without a THz pulse applied on the sample. More features are very relevant for system optimization, and therefore the software includes: 5) a routine for signal optimization, in order to observe a real-time scan of the signal on screen, although with a large noise; 6) a signal-channel monitor, which displays the actual values of each channel of the DAQ before averaging; 7) a Delay Stage control to check for the device performance and to drive it manually and independently on the measurements. We note incidentally that point 6 can be used also as a signal monitor to manually measure the signal as a function of any variable which is not under the software control (temperature, external electric or magnetic fields, and so on...). Since the signal is revealed as the difference between THz-ON and THz-OFF trains of pulses, this reading cannot be done by means of a single external unit, such as a digital oscilloscope.

The software can manage automatically the creation of folders and data files for saving the experimental results and the parameters are checked at each input in order to ensure that any possible mistake by the user cannot create problems during the measurement execution. At every startup the initialization procedure is collecting data from all devices to set the safety breaks which are forcing the user to set all parameters in within a safe range. Moreover every value is rounded to the minimal time/angle/wavelength resolution, so that the device can be driven with no overlap between subsequent points. The measurements files are numbered incrementally in order to ensure that two runs with identical parameters and names will be stored in different files, which are saved automatically during each run, but can be deleted at the end of the run upon user command. The software also has the ability to ring a bell when the measurement is finished, and the bell stops ringing as soon as the mouse is moved or any key is pressed. In this way the user can save time and energy by working on other things while the measurement is running, still without losing any time at the end of each run. The basic idea below TeraVision is that the user must be concentrated uniquely on the measurement and not be bored by many irrelevant tasks.

The software produces three distinct files for each single measurement: a large data file in which every single point of every scan is saved before averaging, a short data file in which only the average on all points and scans is saved, and an image file which contains a summary of the measurement with graphs and all relevant parameters and comments by the user. This ensures that during data analysis the user will find quickly what he/she is searching for, he/she will be able to quickly analyze the averaged data, but he/she will also be able to perform a deep data analysis on individual points if needed, so to maintain the full information content of the measurement. The data files are saved point-by-point during the run, in order to prevent information losses due to system crash, electrical black-out or whatever unpredictable problems.

All these features are meant to create a very friendly and free-of-worries user experience, to make unnecessary for the user to have any specific knowledge about the software code and to achieve 100% safety about data loss and device operation.

From the point of view of the performances, and in particular in order to reduce the measurement time as much as possible, several solutions have been implemented, and they will be discussed in the next sections.

3.3 The TeraVision front panel

We begin to describe the operation of the TeraVision software starting from the final user interface, in order to explain in some details all the features included in the software. In a second moment, we will describe the "backstage", i.e. how we implemented the proposed solutions in the software code. **Figure 5** shows the TeraVision front panel. In the upper line of the panel there is the software name and version, and the Quit button. This button is important, for it drives the correct quitting procedure of the software, as better explained in the dedicated section. The user can quit the software in other ways, but then he/she will need to close manually all ports used by the program and all set parameters of that experimental session will be lost and the software will load the previously saved ones at next startup. Below the first line, we can locate two main blocks: the central Tab panel, which is

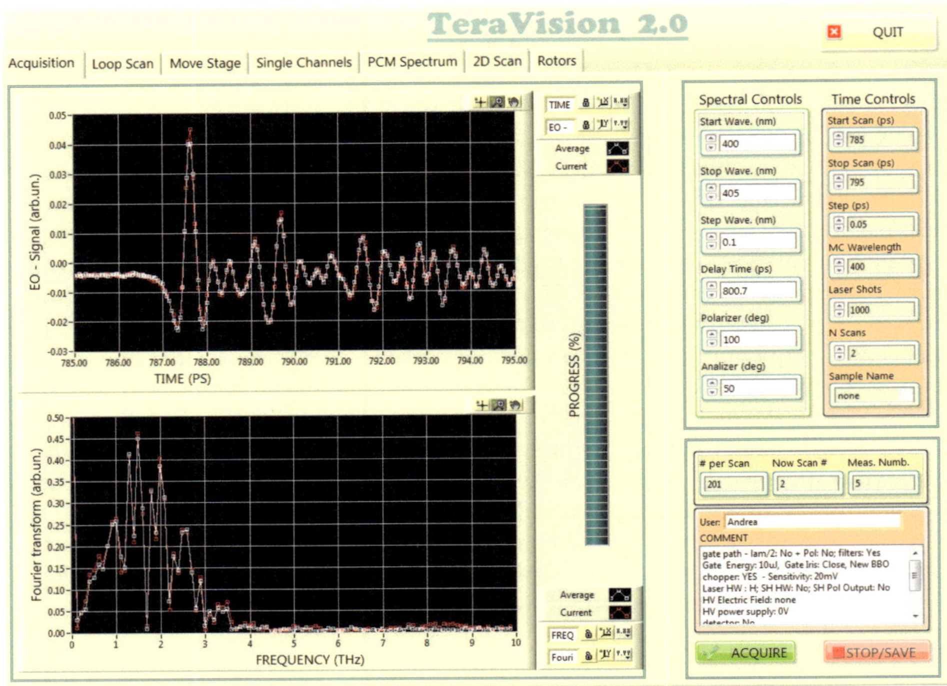

Figure 5.
TeraVision front panel.

the core of the software, and a lateral column. This column will be always on screen for any Tab selected. Let us describe the column content, and then we will describe each individual Tab of the central panel in separate subsections.

3.3.1 The controls column

The lateral column is always on screen, for each Tab selected. It is composed of two upper Controls columns and one lower indicator/comments column. On the two upper columns the user can set the parameters for different types of measurements. These two columns have been labeled "Spectral controls" and "Time controls" not because they contain only spectral or time controls (they do not) but rather to remind the user that when he/she would like to perform a spectral measurement, the software will drive the Delay Line to a specific time (so the Delay Time parameter is shown in the Spectral Controls). Similarly, when the user wants to perform a Time-scan (just as the one shown in **Figure 5**) the software will set the monochromator wavelength to a fixed specific value, and therefore this value has been inserted in the Time Controls column. In the lateral columns it is possible to set the parameters for any specific measurement. The user can define the start/stop/step values for a time scan or a wavelength scan, the angles of polarizer and analyzer waveplates, the number of Laser shots on which the average will be performed and the number of Scans (the number of times the single measurement will be repeated to improve the signal-to-noise ratio) and finally the sample name which will be used in the experimental report. Every single value is checked together with the others in order to ensure safety. For instance if the user sets by mistake a value for stopping the Delay Line which is larger than the Delay Line length itself, the software force the value to the maximum, which has been recorded during the device initialization procedure. If the start value is larger than the stop value, the two values are exchanged. If the user sets a negative value for the Laser Shots parameter, it is forced to positive, and so on.

3.3.2 The comments columns, the saved files and the experimental report

Below the Controls columns there are three indicators. "# per Scan" indicates the number of points to measure within a single scan, "Now scan #" indicates the number of current scan and "Meas. Numb." indicates the incremental measurements numbering. As already mentioned, each time a measurement is launched, the software creates three files using the sample name, date/time and measurement number for naming the files. Those files are saved in a folder, which is created automatically at each startup during initialization, named with the date of the measurement. The three files have all identical names but different file extensions. The *.dat file contains all single points of the measurement. The *.txt file contains only the averages for all points and all scans. The third file is a *.jpeg file with the printout of the experimental report.

Below the indicators there are two text fields with the user name (for future references) and a free textbox in which the user will write all parameters which are not under TeraVision control (sample temperature, electric or magnetic fields applied to the sample, and so on...) and whatever observation is relevant to the present experiment. This "comment" textbox is extremely useful for creating the experimental report file. **Figure 6** shows an example of a report file. The measurement number and sample name are in the top row of the page. In the center of the page there are the graphs of the measurement, and just below them there is a textbox with all relevant parameters, the units of each column in the data files and the free comment textbox, just described above. The reason while we have chosen the *.jpeg format instead of the more used *.pdf file is that having one folder with up to 100 or more measurements, the user must open those 100 *.pdf files individually in order to find the measurement he/she is searching for. Using jpeg and any picture viewer, one can just browse all pictures in the folder within the same window simply by pressing an arrow key, and it is even possible to use the preview function so that one does not need to actually open the file in order to know what is inside that specific measurement. We found that if one makes use of this report wisely and carefully, it can create a digital log-book which can partially replace the need of a physical hand-written lab-book.

3.3.3 The Tab panel

- The first Tab is shown in **Figure 5** is the "Acquisition" Tab. It is used for time-scan, both with electro-optic detection, as in regular THz-TDS, or for scanning in time on a single wavelength in a THYR experiment, or for THz-TDS with alternative techniques such as the already mentioned ABCD technique. There are two graphs, one for the measured signal in time domain and the other which displays the Fast Fourier Transform (FFT) of the signal. In each graph two lines with different colors are displayed. The red one is the measurement on the current scan, and it is updated point-by-point, so that the user can immediately see if there are problems during the measurement without need to wait one full scan to be finished. The white curve is the average of all previous scans, and it is updated at the end of each scan. In the saved txt data file and jpeg report only the averaged curve is stored. In addition, the user has all the usual LabVIEW graph tools for changing the scale, labels, and more. Finally a vertical bar is tracing the progress of the measurement. This progress is not a theoretical estimate. It is calculated by measuring the real elapsed time since

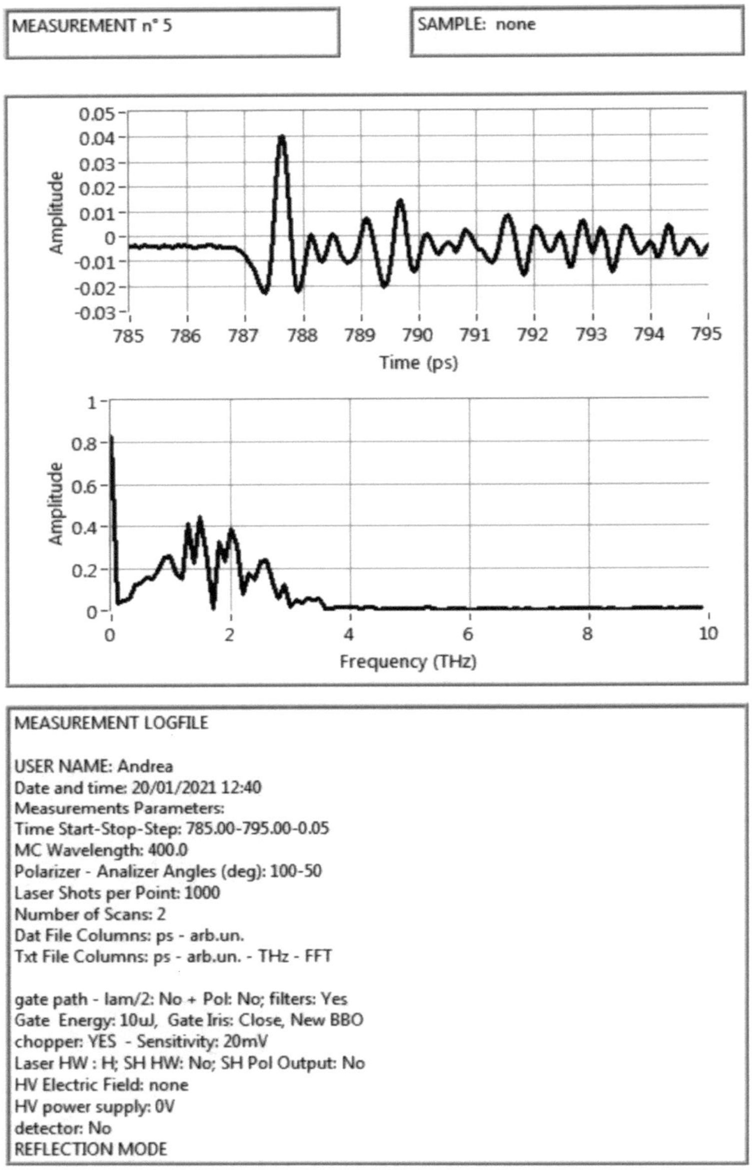

Figure 6.
The report jpeg file generated at the end of each experiment.

the beginning of the measurement, dividing this time by the number of measured points so to obtain a realistic estimate of the time-per-point, and finally a multiplication for the number of total points and remaining points gives an estimate of the overall progress, in percent.

- The second Tab is the "Loop Scan" Tab. In this Tab the time scan is shown with no averaging. The delay stage is moved continuously in between the start and stop point, and the positions are not recorded. This speeds up a lot the acquisition, so that the waveform appears on the PC screen in real-time exactly as if it was displayed on an oscilloscope screen. In this Tab it is possible to set the velocity and acceleration of the Delay Stage, which will be then used also elsewhere. The reason for this utility is to monitor the signal while adjusting the alignment of the setup. Who works with THz knows that the setup optimization can be quite difficult because if one looks at some signal on the oscilloscope, for instance the peak of the waveform, this peak position will change in time as the user is optimizing the alignment, because the relative path between THz and Gate pulses is changed. If one has the signal on screen in real-time, the displacement of the peak position will not be a problem, as long as the peak will fall between the start and stop points. The major drawback of this method is that the very poor signal-to-noise ratio makes difficult to distinguish the signal when it is very far from optimal intensity. It is very useful to quickly reach the optimum once the signal is already roughly optimized.

- The third Tab is the "Move Stage" Tab. Here the user can set the Delay Stage parameters, and it displays a real-time chart which shows the stage position in time. This utility is used to manually move the stage step-by-step when searching for the signal after a major realignment, or to quickly move the stage in between two fixed positions. For instance, when a filter is placed in one of the two arms THz or Gate, it will change the relative optical path by a fixed known amount of time, so that it will be possible to quickly switch between the peak-signals in the two different cases.

- The fourth Tab is the "Single Channels" Tab. In this Tab a graph in real time shows just the output of the DAQ for a given number of points. The DAQ receives two signals. One is the photodetector (PMT or balanced photodiode), and the second channel is linked to a photodiode which monitors the THz generation optical pulse right after the chopper. This monitor is needed for two reasons: the first is to keep trace of the laser intensity, and the second to know exactly which pulse has been chopped and which one has not. Even if in principle this signal could be used to normalize the measured signal on channel 1 in order to account for laser intensity fluctuations, our test measurements reveal that the amount of noise introduced by this procedure greatly overcomes the benefits of having a normalized signal, and therefore the channel 2 is used only to recognize the ON and OFF signals so to perform the differential measurement. In this Tab it is possible to set the DAQ parameters such as number of triggers to wait and average for each experimental point (delay stage position, monochromator wavelength). The utility is also capable of warning the user if the channel 2 photodiode signal difference between ON and OFF pulses is lower than 3 times the sum of the two standard deviations of each series of points (ON and OFF). When this condition is met, it means that the chopper phase has somehow changed and the chopping does not completely block every second pulses (this may happen for instance when the laser trigger has significant jitter or when the trigger parameters in the chopper motor are not correctly set). In this case, a red indicator blinks and the user is invited to check the chopper phase.

- The fifth Tab is the "PCM Spectrum" Tab. Here the signal is measured as a function of wavelength for a fixed Delay Line position (time position). In this Tab it is possible to set the monochromator parameters and to record a curve of the signal when THz is ON, OFF or the difference between the two. This is important in the case of THYR measurements, in order to directly compare the signal with no THz (standard SHG signal) and the signal with THz applied to the sample (THYR signal). It is also quite useful for quickly checking the presence of spurious effects, as for instance 2-photons luminescence.

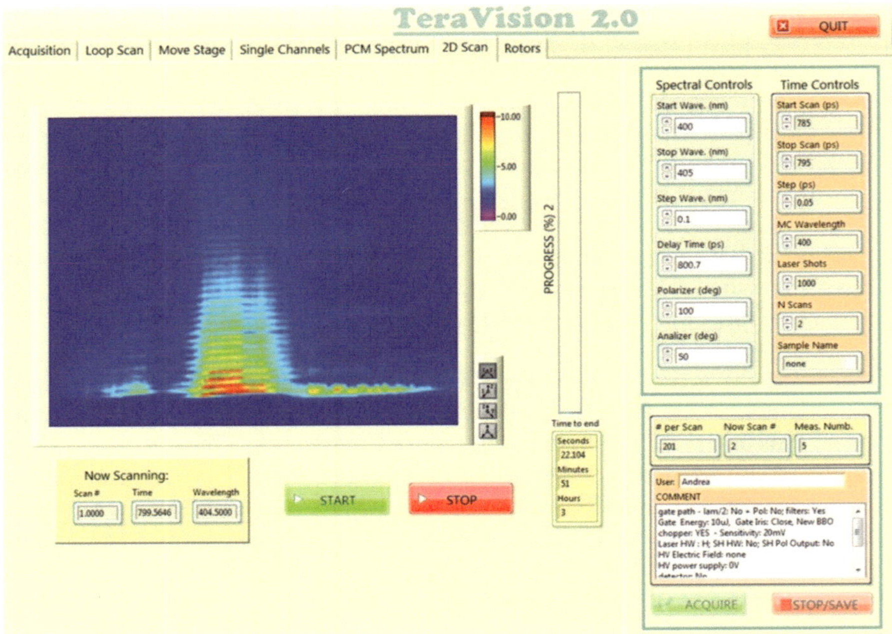

Figure 7.
2D scan tab for THYR spectroscopy.

- The sixth Tab is the "2D Scan" Tab. This Tab is shown in **Figure 7**. This is the Tab for THYR measurement, where the signal is measured in 2D as a function of both wavelength (x-axis) and time (y-axis). The graphs in this Tab can be manipulated to be a color mesh in 2D or a full 3D graph. The figure shows a typical THz measurement, in which we can observe first and second order Stokes and anti-Stokes bands and their very different time relaxation. The oscillating behavior of the central peaks is evident as well. For a detailed explanation of the measurement shown in this figure, we would like to refer to Ref. [26].

- Finally, the last Tab is the "Rotors" Tab, in which the signal can be measured as a function of the incoming or outcoming optical polarization. This is of particular importance when studying the symmetry properties of the SHG signal generated by the sample. This is in turn important for obtaining a correct physical interpretation of the THYR signal. This is in fact the result of a complex nonlinear optical effect, and its relationship with the target material parameters of interest can be not easy or immediate to appreciate. One has to usually study the SHG symmetry and characteristics, with and without applying a THz pulse, in order to find the most convenient choice of the optical polarizations. This Tab can be also used as stand-alone acquisition software when a SHG measurement, rather than THYR or THz-TDS, is the real goal of the experiment.

3.4 Initialization and quitting routines

The conceptual scheme of the experiment in **Figure 8** shows the optical, electrical and digital connections of the THYR setup. The TeraVision software controls 4 items: Delay Stage, Data Acquisition Card (DAQ), Rotors and the monochromator. The first task of the software at startup is therefore to initialize all these devices.

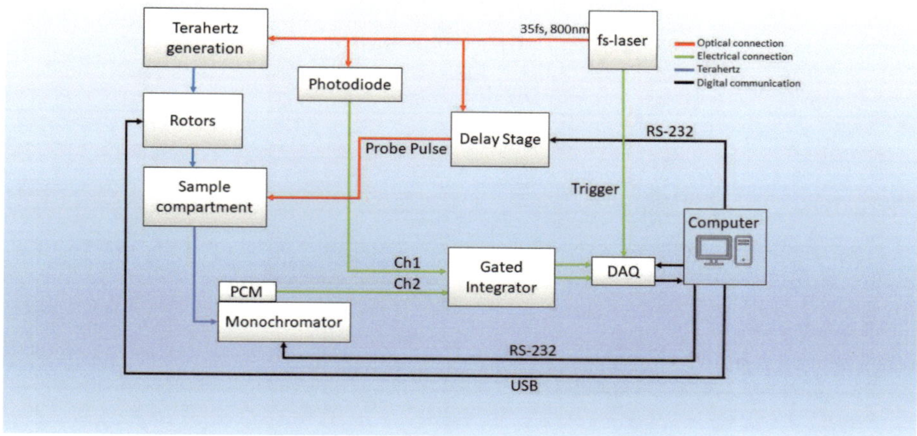

Figure 8.
Conceptual scheme of the experiment.

The program creates all folders needed for data acquisition according to the system date and time; it opens all communication ports and cleans all buffers. In this phase, however, the software is not only performing the basic initialization in order to be able to communicate with each device, but it is also interacting with the user. Using pop-up messages the software asks the user about starting set parameters, which initial value is read in a setup file. For instance the user can choose to move the Delay Stage to find its physical edges or to set the "home position", i.e. the zero-step position. The user can choose to skip the initialization procedure of that specific device too. The same happens for all devices. This start-up phase is run only at the beginning, when the software is launched. Then the software is running in time-out mode, waiting for user commands. When the software is quit, a second one-time-only routine is launched, which is saving all set parameters in the setup file in order to call them again at next startup. This method is very convenient as the parameters are many, and the user can not always remember which parameter was set last time. Beside this, the quit-routine is closing all open ports and flushing all memories of the devices in order to ensure a clean startup on the next call. A "smart-stop" VI ensures that the user cannot quit the program if one or more while/for loops are still running. The VI makes sure all loops are closed in the right order, from internal to external, and all queues are correctly flushed before moving to stop the next loop. This is the reason why it is not recommendable to quit the software in other ways.

3.5 The block diagram: event structure/producer and consumer loops

After initialization, the software goes in timeout mode and waits for user commands. The entire architecture of the software is therefore enclosed in a large event structure, shown in **Figure 9**, where every single button on screen is triggering a specific event. This choice of the basic architecture was perhaps not the wisest one, and it will be changed in the future (see the final section about future perspectives). In order to ensure that unjustified clicking on certain buttons by the user can harm the software execution, each event will disable all buttons which are not relevant during the performance of that event, and those buttons will be grayed, so that the user has an exact feeling of what can and cannot be done in that precise moment. The user can always stop the current task with a specific stop button, which drives

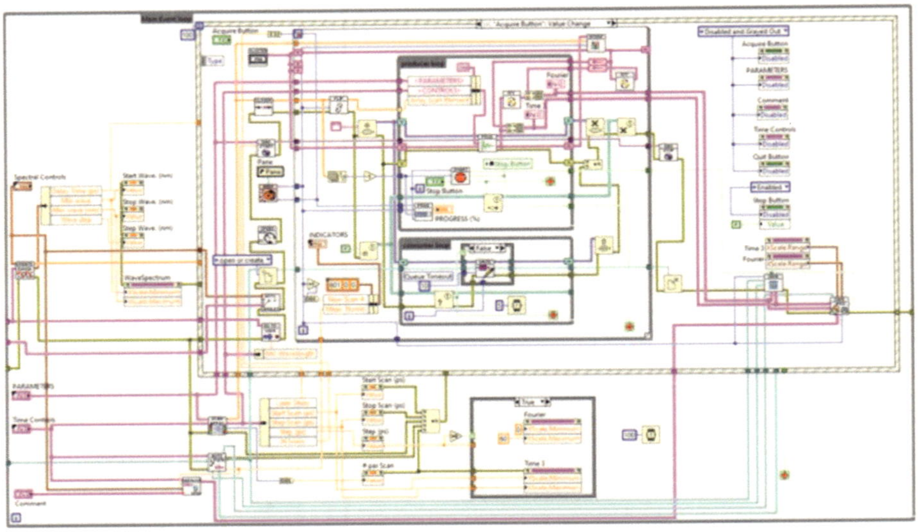

Figure 9.
The TeraVision block diagram.

the already mentioned "smart-stop" VI instead of being directly linked to any loop, so to ensure that the system has returned to timeout mode correctly.

The TeraVision software has been written with particular effort on saving time during each measurement. If a simple time-scan on a short range and few points can take seconds or few minutes, a full resolution 2D scan of a real THYR measurement can take several hours. This time is really at the limit of long-term laser stability specifications. For this reason, every millisecond spared during the single point acquisition, or in between two subsequent scans, is important. Several actions have been taken in order to reduce the measurement time as much as possible, and we will list them in this section.

First, the time required for the actual measurement (devices movements and number of triggers for each point) is decoupled from all analysis and digital work (in particular data and parameters file writing, which is the most digital time-consuming task). The way we chose for this, is to implement several Producer/Consumer loops (an example is shown in **Figure 10**). The Producer/Consumer architecture is a classic case of multithreaded synchronization solution. It applies when two threads (producer and consumer) actually run together in parallel. The main role of the producer is to generate a certain amount of data to put into the buffers and then repeat the process with one or more nested for/while loops. It is crucial to implement "smart stops" in order to prevent crashes due to nested loops. At the same time, consumers are consuming the data in buffers for data analysis and saving. The key is to ensure that producers do not add data when the buffer is full, and consumers do not consume data when the buffer is empty. **Figure 10** shows as an example the Producer/Consumer architecture for the Acquisition task. Many users know that LabVIEW can be programmed by multithreading, but usual ways to implement multithreading are to use global and local variables, with the disadvantages of possible data loss, risk and competition. The loops are driven by queue controls in order to ensure a correct processing of data flow.

The data processing is, however, not the most time-expensive problem. The most time-consuming tasks are by far the physical movements of all devices. In order to minimize those movements, we implemented the following procedure.

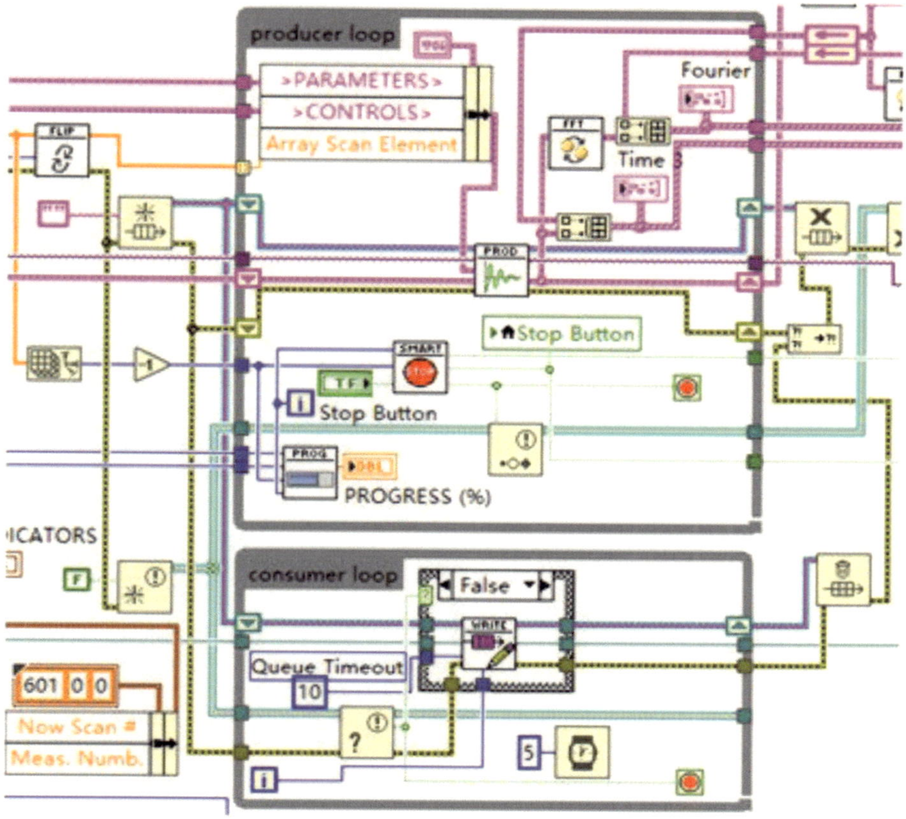

Figure 10.
Producer-consumer loop.

The usual way to drive a step motor is to give it a target position, then read out the actually reached position and finally adjust it with a secondary movement command, with or without a third step due to so-called backlash correction. We chose to avoid all position corrections. For the measurement is not important that the target position is "really" that given in command, as long as the actual position reached after a single movement is read with sufficient precision and accuracy. Therefore, the system creates an array of positions, from the start/stop/step parameters, equally spaced, and gives them to the motor driver VI for each trigger of the acquisition loop. Instead of waiting that the device goes precisely on each position of the array, the software collects data on random positions acquired "closed to" the target positions. After one scan is completed, the software interpolates the signal values acquired on random positions on top of the original equally spaced array. In this way we realize two important things: first, the signal can be averaged easily with all previous scans as the x-axis values are identical, and second the FFT routine will work correctly, as it requires to run over equally spaced arrays only. For the same reason of minimizing the device movements, the software is also capable of performing the scans in "reverse" mode, i.e. it scans from start to stop and then from stop to start, in order to remove the dead time needed to send the motor back to the start position. Finally, the system can decide automatically how to start the first scan by reading the actual position and compare it with the position array so to choose which direction of motion will give the shortest time before the measurement begin.

4. Future perspectives

In the near future we plan to upgrade the software to version 3.0 with three main improvements (and many minor adjustments). First, the system will be able to perform individual separated measurements, saved in separated individual files, for a given list of static parameters. For example, we can now scan the signal as a function of the input polarization, and we can then fix one specific angle and run a time-scan or a wavelength-scan. In the future it will be possible to programmatically set two or more angles (and other parameters as well) and run subsequent separate measurements for every single parameter.

The second improvement goes in the direction of greatly reducing the measurement time. We bought a new Delay Line which is capable of reading the stage position on-the-fly while the stage moves. Those positions are recorded in a buffer and they can be read all together in the dead-time which the motor employs to stop the stage and start to scan in the opposite direction. The system can therefore work continuously without the need to make stop-and-go movements. The challenge here is represented by the need to link the physical position of the stage in real time, to the time at which the signal is actually measured, and this requires a very fine and precise synchronization of the digital clock with the laser trigger.

Finally, a major rearrangement of the whole software architecture is needed in order to bring the loops out of the event structure. A simple state machine will contain the measurement loops, while the event structure will drive the state machine. This rearrangement is not critical because our current architecture has proved itself to be very robust (the software crashes very rarely, if ever), but we have in mind to do so because it will be much easier to work on the software when it is necessary to add/change any component, and also in order to have a much more "clean" block diagram.

Acknowledgements

The authors would like to acknowledge funding from the Italian Government, PRIN-TWEET (grant code 2017YCTB59).

Author details

Rohit Kumar[1], Qiucheng Yu[1], Domenico Paparo[2] and Andrea Rubano[1*]

1 Dipartimento di Fisica "E. Pancini", Università Federico II, Napoli, Italy

2 CNR-ISASI, Institute of Applied Sciences and Intelligent Systems, Consiglio Nazionale delle Ricerche, Pozzuoli, NA, Italy

*Address all correspondence to: andrea.rubano@unina.it

References

[1] Lewis R. A. A review of terahertz sources, J. Phys. D: Appl. Phys. 2014; 47: 374001. DOI:10.1088/0022-3727/47/37/374001.

[2] Spies JA, Neu J, Tayvah UT, Capobianco MD, Pattengale B, Ostresh S, et al. Terahertz Spectroscopy of Emerging Materials. J. Phys. Chem. C. 2020;**124**:22335-22346. DOI: 10.1021/10.1021/acs.jpcc.0c06344

[3] Jeon T, Grischkowsky D. Nature of Conduction in Doped Silicon. Phys. Rev. Lett. 1997;**78**:1106-1109. DOI: 10.1103/PhysRevLett.78.1106

[4] Kojima S, Tsumura N, Wada Takeda M, Nishizawa S. Far-infrared phonon-polariton dispersion probed by terahertz time domain spectroscopy. Phys. Rev. B. 2003;**67**:035102. DOI: 10.1103/PhysRevB.67.035102

[5] Nicoletti D, Cavalleri A. Nonlinear light-matter interaction at terahertz frequencies. Adv. Opt. Photonics. 2016;**8**(3):401-464. DOI: 10.1364/AOP.8.000401

[6] Arjavalingam G, Theophilou N, Pastol Y, Kopcsay GV, Angelopoulos M. Anisotropic conductivity in stretch-oriented polymers measured with coherent microwave transient spectroscopy. J. Chem. Phys. 1990;**93**: 6-9. DOI: 10.1063/1.459520

[7] Robertson WM, Arjavalingam G, Maede RD, Brommer KD, Rappe AM, Joannopoulos JD. Measurement of photonic band structure in a two-dimensional periodic dielectric array. Phys. Rev. Lett. 1992;**68**:2023-2026. DOI: 10.1063/1.459520

[8] Thrane L, Jacobsen RH, Uhd Jepsen P, Keiding R. THz reflection spectroscopy of liquid water. Chem. Phys. Lett. 1995;**240**:330-333. DOI: 10.1016/0009-2614(95)00543-D

[9] Mou S, Rubano A, Paparo D. Complex Permittivity of Ionic Liquid Mixtures Investigated by Terahertz Time-Domain Spectroscopy. J. Phys. Chem. B. 2017;**121**(30):7351-7358. DOI: 10.1021/acs.jpcb.7b04706

[10] Mou S, Rubano A, Paparo D. Broadband Terahertz Spectroscopy of Imidazolium-Based Ionic Liquids. J. Phys. Chem. B. 2018;**122**(12):3133-3140. DOI: 10.1021/acs.jpcb.7b10886

[11] HsiehY D, Nakamura S, Abdelsalam DG, Minamikawa T, Mizutani Y, Yamamoto H, et al. Dynamic terahertz spectroscopy of gas molecules mixed with unwanted aerosol under atmospheric pressure using fibre-based asynchronousoptical-sampling terahertz time domain spectroscopy. Sci. Rep. 2016;**6**:28114. DOI: 10.1038/srep28114

[12] Tielrooij KJ, Paparo D, Piatkowski L, Bakker HJ, Bonn M. Dielectric relaxation dynamics of water in model membranes probed by terahertz spectroscopy. Biophys. J. 2009;**97**:2484-2492. DOI: 10.1016/j.bpj.2009.08.024

[13] Haddad EJ, Bousquet B, Canioni L, Mounaix P. Review in terahertz spectral analysis. Trends. Analyt. Chem. 2013;**44**:98-105. DOI: 10.1016/j.trac.2012.11.009

[14] Davies AG, Linfield EH, Johnston MB. The development of terahertz sources and their applications. Phys. Med. Biol. 2002;**47**:3679-3689. DOI: 10.1088/0031-9155/47/21/302

[15] Song HJ, Nagatsuma T. Present and Future of Terahertz Communications. IEEE. 2011;**1**:256-263. DOI: 10.1109/TTHZ.2011.2159552

[16] Iwaszczuk K, Heiselberg H, Jepsen PU. Terahertz radar cross section

measurements. Opt. Express. 2010;**18**:26399-26408. DOI: 10.1364/OE.18.026399

[17] Kulesa C. Terahertz Spectroscopy for Astronomy: From Comets to Cosmology. IEEE. 2011;**1**:230-240. DOI: 10.1109/TTHZ.2011.2159648

[18] Lu MH, Shen JL, Li N, Zhang Y, Zhang CL. Detection and identification of illicit drugs using terahertz imaging. J. Appl. Phys. 2006;**100**:103104. DOI: 10.1063/1.2388041

[19] Masanori H. Development and future prospects of terahertz technology. Jpn. J. Appl. Phys. 2015; 54:120101. DOI: 1347-4065/54/12/120101

[20] Schmuttenmaer CA. Exploring Dynamics in the Far-Infrared with Terahertz Spectroscopy. Chem. Rev. 2004;**104**:1759-1779. DOI: 10.1021/cr020685g

[21] Strickland D, Mourou G. Compression of amplified chirped optical pulses. Opt. Commun. 1985;**56**:447-449. DOI: 10.1016/0030-4018(85)90151-8

[22] Kampfrath T., Tanaka K., Nelson K. A. Resonant and nonresonant control over matter and light by intense terahertz transients. Nat. Photonics2013; 7: 680-690. DOI: 10.1038/NPHOTON. 2013.184

[23] Khiabani N, Huang Y, Shen YC, Stephen J. Theoretical Modeling of a Photo-conductive Antenna in a Terahertz Pulsed System. IEEE Trans. Antennas Propag. 2013;**61**:1538-1546. DOI: 10.1109/TAP.2013.2239599

[24] Zhang XC, Ma XF, Jin Y, Lu TM, Boden EP. Terahertz optical rectification from a nonlinear organic crystal. Appl. Phys. Lett. 1992;**61**:3080-3082. DOI: 10.1063/1.107968

[25] Zhong H, Karpowicz N, Zhang XC. Terahertz emission profile from laser-induced air plasma. Appl. Phys. Lett. 2006;**88**:261103. DOI: 10.1063/1.2216025

[26] Rubano A, Mou S, Marrucci L, Paparo D. Terahertz Hyper-Raman Time-Domain Spectroscopy. ACS Photonics. 2019;**6**:1515-1523. DOI: 10.1021/acsphotonics.9b00265

[27] Wang Q, Ma Y. Qualitative and quantitative identification of nitrofen in terahertz region. Chemometr. Intell. Lab. Syst. 2013;**127**:43-48. DOI: 10.1016/j.chemolab.2013.05.011

[28] Amoruso S, Andreone A, Bellucci A, Koral C, Girolami M, Mastellone M, et al. All-carbon THz components based on laser-treated diamond. Carbon. 2020;**163**:197-201. DOI: 10.1016/j.carbon.2020.03.023

[29] Walther M, Plochocka P, Fischer B, Helm H, Jepsen PU. Collective vibrational modes in biological molecules investigated by terahertz time-domain spectroscopy. Biopolymers. 2002;**67**:310-313. DOI: 10.1002/bip.10106

[30] Kaindl RA, Carnahan MA, Orenstein J, Chemla DS, Christen HM, Zhai HY, et al. Far-Infrared Optical Conductivity Gap in Superconducting MgB_2 Films. Phys. Rev. Lett. 2002;**88**:027003. DOI: 10.1103/PhysRevLett.88.027003

[31] Rini M, Tobey R, Dean N, Itatani J, Tomioka Y, Tokura Y, et al. Control of the electronic phase of a manganite by mode-selective vibrational excitation. Nature. 2007;**449**:72-74. DOI: 10.1038/nature06119

[32] Jepsen PU, Fischer BM. Dynamic range in terahertz time-domain transmission and reflection spectroscopy. Opt. Lett. 2005;**30**:29-31. DOI: 10.1364/OL.30.000029

[33] Mou S, Rubano A, Paparo D. Terahertz hyper-Raman time-domain spectroscopy of gallium selenide and its application in terahertz detection. Appl. Phys. Lett. 2019;**115**:211105. DOI: 10.1063/1.5115986

[34] Ceraso A, Mou S, Rubano A, Paparo D. Coherent THz Hyper-Raman: Spectroscopy and Application in THz Detection. Materials. 2019;**12**:3870. DOI: 10.3390/ma12233870

[35] Xie X, Dai JM, Zhang XC. Coherent Control of THz Wave Generation in Ambient Air. Phys. Rev. Lett. 2006;**96**:075005. DOI: 10.1103/PhysRevLett.96.075005

[36] Rubano A, Braun L, Wolf M, Kampfrath T. Mid-infrared time-domain ellipsometry: Application to Nb-doped SrTiO3. Appl. Phys. Lett. 2012;**101**:081103. DOI: 10.1063/1.4746263

[37] Mou S, Rubano A, Paparo D. A new THz-Pump/Second Harmonic Probe experimental setup: characterizations and first results. 2016;18th Italian National Conference on Photonic Technologies.

Chapter 3

Cost-Effective Interfaces with Arduino-LabVIEW for an IOT-Based Remote Monitoring Application

Muhammad Asraf Hairuddin, Nur Dalila Khirul Ashar, Amar Faiz Zainal Abidin and Nooritawati Md Tahir

Abstract

To date, research efforts have demonstrated the stimulated need for the Internet of Things (IoT) based monitoring device in their laboratory. The benefits of remote laboratories in overcoming time constraints and the disadvantages of usability of conventional laboratories are well known. In addition to the current control engineering laboratories, a remote lab that incorporates an industry-relevant method has been established to assist in the understanding of data acquisition with cost-effective platform integration. However, one of the greatest challenges is the creation of a low-cost and user-friendly remote laboratory experiment that is ideal for interacting with the actual laboratory via a mobile device. The main objective of this work is therefore to build a remote laboratory system based on the IoT using the LabVIEW-Arduino interface with the example of proportional-integral-derivative (PID) tuning scheme for the LD-Didactic temperature plant. The practical work would include the implementation of the low-cost Arduino module connecting the actual plant to mobile devices. In addition, interfaces have been built using the Blynk application to allow communication between the end user and the laboratory equipment. In line with the Industrial Revolution 4.0 (IR 4.0), the proposed study structure called for the digitization of the current laboratory experiment method.

Keywords: Arduino-LABVIEW, temperature application, remote laboratory, interfacing module, data acquisition, internet-of-things

1. Introduction

Studies in remote laboratory are one form of teaching adaptation that reflects the delivery of practical experience in learning. The common transition from traditional laboratory to remote laboratory has always been associated with automation and online technology, while common implementation involves embedded or monitoring related experiments. As a result, an interactive laboratory can be developed to engage students' comprehension in the learning process. It is worth to shift the laboratory implementation without overlooking the hands-on element. This is in line with the advancement of IR4.0 which extends the education system with IoT into endless possibilities in online learning.

This would support both students and instructors by enhancing the learning environment and exposing students to the industrialized sector. As a core learning process, laboratory experience is essential to learning processes in all areas of engineering. Studies have shown that students who participate in well-designed laboratory experiences gain valuable skills and competencies. Some agree that the development of laboratory activities that enable students to conduct experiments remotely would increase students' commitment to furthering their education. At the same time provide training for high-tech careers by fostering the skills desired by potential employers [1]. Indeed, this not only allow students to actively participate in the learning process, but they will also have vivid experiences, can work in a group or independently, and their attention will improve their commitment and satisfaction [2]. On the management side, remote laboratories benefited by reducing the number of scheduling arrangements, overcoming equipment inadequacy, and reduce overloading students by utilizing the learning versatility provided by IoT.

As a result, the need to transform the traditional laboratory into a remote laboratory is significant. In these circumstances, effective interfaces must be established in order to achieve system monitoring as well as remote application interconnectivity through the internet. Such requirements can be met by using a data acquisition system, in this case LabVIEW (Laboratory Virtual Instrument Engineering Workbench) in conjunction with Arduino board. Furthermore, the proposed system is not limited to laboratory applications and can be extended to other applications such as solar photovoltaic, agriculture, and environmental studies. The importance of such a framework can be seen in open-source hardware as a data acquisition device that is easily configurable to meet the intended requirements of specific applications.

The remaining chapter discusses the trend and review on the open-source interfacing module for remote applications along with implementation challenges in Section 2. Section 3 presents the methodology, to cover overall software and hardware used in this work implementation of interfacing module to acquire the data. The process known as data acquisition will demonstrate how the data transferring process is designed to collect the data from and to the real plant in the laboratory. Section 4 and Section 5 discuss Blynk integration to modular hardware as well with example of case study for PID simulation in real time execution.

2. Related works and research trends

2.1 Trend analysis for remote monitoring

The remote laboratory has been used for the past 20 years and is still used progressively for learning technologies, as shown by the publication pattern depicted in **Figure 1**. A basic bibliometric analysis was performed, to compile and evaluate the information obtained from the SCOPUS database's bibliographic sources. This analysis includes a summary of the most relevant keywords related to monitoring applications, as well as the publications trend and subject area that have been published. Over the last five years, publications with an average of more than 50 articles revealed that research interest was concentrated on these keywords namely monitoring, cost, architecture, e-learning, and higher education, as shown in **Figure 2**. The state of the art has presented the most recent status of remote technology progresses, implementations, and applications through many niche studies. This study offers insight into discovering the most important publications of the year, how interest in such subjects has grown over time, and the most studied topics in the subject area.

Current trends in total publications, as shown in **Table 1**, indicate that this research is expected to receive demand from the engineering and education sectors

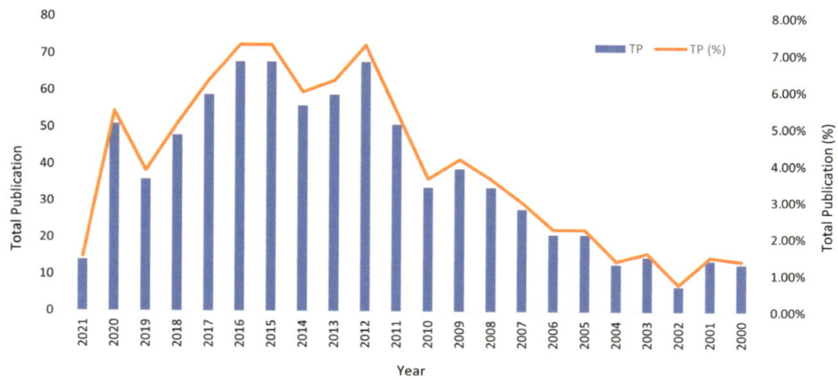

Figure 1.
Total publications by year.

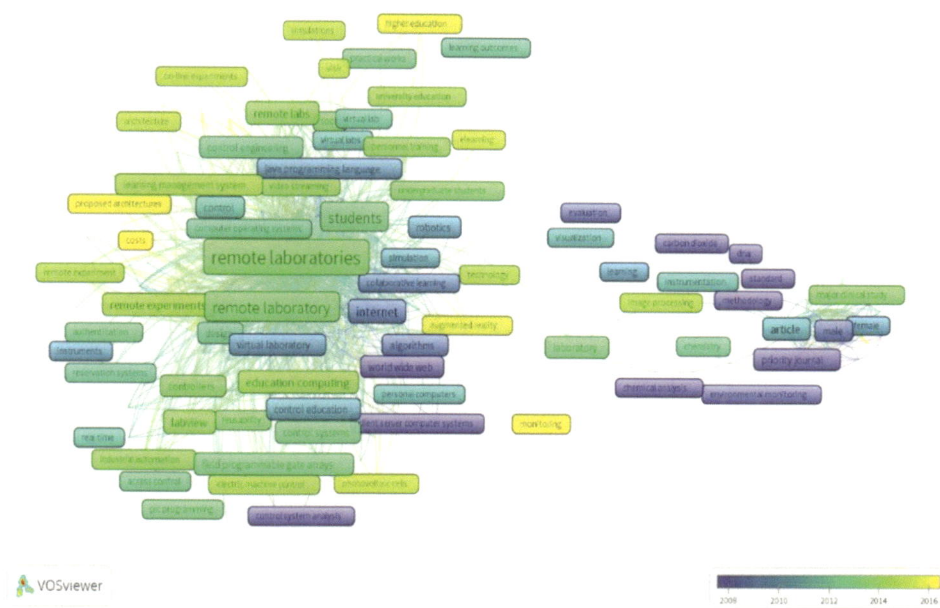

Figure 2.
Top keywords with VOSViewer visualization map.

to establish remote monitoring. Surprisingly, engineering accounts for 53% of total publications, led by social science, physics and astronomy, and computer science, which each contribute between 27%, 25%, and 22% in the subject field of remote monitoring applications.

2.2 Consideration of LABVIEW in remote laboratory application

LABVIEW offers simple interfaces in form of graphical programming through Graphical User Interface (GUI) instead of text-based programming. The environment development of the project with front panel such as controls (known as input) to supply information to the VI, indicators (known as output) display the results based on the inputs given to the VI and block diagram to comprise of graphical block programming that applies data flow concept is known as virtual instruments (VIs).

Subject Area	TP	%
Engineering	501	53%
Social Sciences	252	27%
Physics and Astronomy	232	25%
Computer Science	203	22%
Chemistry	43	5%
Medicine	39	4%
Mathematics	29	3%
Biochemistry, Genetics and Molecular Biology	28	3%
Agricultural and Biological Sciences	24	3%
Materials Science	23	2%
Environmental Science	19	2%
Earth and Planetary Sciences	17	2%
Energy	17	2%
Business, Management and Accounting	13	1%
Chemical Engineering	12	1%
Psychology	10	1%
Arts and Humanities	9	1%
Immunology and Microbiology	8	1%
Neuroscience	5	1%
Pharmacology, Toxicology and Pharmaceutics	5	1%
Veterinary	5	1%
Multidisciplinary	4	0%
Decision Sciences	3	0%
Dentistry	3	0%
Economics, Econometrics and Finance	3	0%
Health Professions	3	0%

Table 1.
Subject area.

Dataflow programming in LABVIEW executes the flow of data through the nodes on the block diagram in sequence order. The design of VIs will determine the block diagram structure based on function code, in which data flow through the interconnected wires. When all of its inputs are available to execute the function code, it supplies data to its output terminals and passes the output data to the next node in the dataflow direction. Most other text-based programming languages are executed by a block diagram node and adopt a data flow model of programme execution. The execution order of a programme is determined by the sequential order of programme elements in control flow. All the items on the front panel will appear as terminals on the back panel. The virtual Instrument Software Architecture (VISA) is used to interface standard I/O for instrumentation programming. To summarize, **Figure 3** depicts the key components as mentioned above that will be implemented as a remote monitoring system.

Various other studies of remote laboratory applications have been applied, such as in the field of physics [3], electronic sensors [4], vibrating beam [5], and rain gauge [6]. Previous researchers demonstrated various type of innovation in their remote learning. Galan D. et.al [7] have demonstrated the usefulness of conducting remote lab experiment for optical levitation that require proper setup to avoid harmful effect on the skin and eyes. Abreu P. et.al [8] have demonstrated the feasibility of

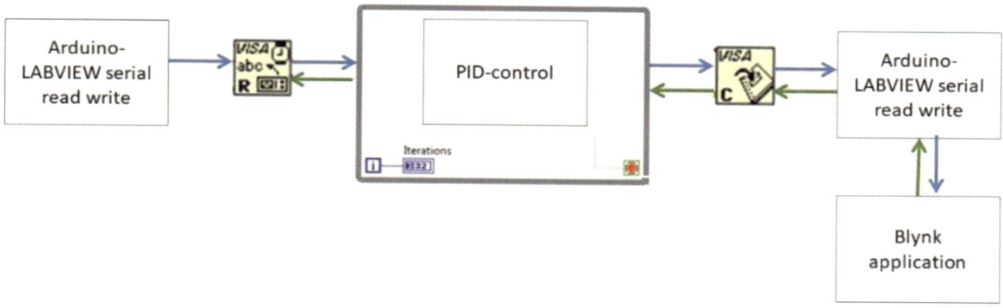

Figure 3.
General visualization block for software dataflow in remote monitoring.

performing experiments and monitoring pressure parameters of a pneumatic system that comprises of valve and pneumatic cylinder. Moreover, other study on robotics application have also ventured into remote laboratory application. Such application was explored by Angulo I. [9] who successfully replicates actual robotics experiment remotely. Based on these literatures, to the best of author knowledge, similar replication in experimenting a temperature process that simulates industrial lab equipment has yet to be developed. The framework replication can be possibly addressed, nevertheless more study is required by considering factors such as number of parameters, existing equipment and the system integration with IOT platform. The integration of temperature process from the laboratory plant to the user device usually incurs high cost when outsourced to third party. **Table 2** summarizes the interfacing technique and data acquisition system that implemented in the remote monitoring development.

This study mainly targeted on the framework development to integration the Arduino-LABVIEW by presenting proof-of-concept implementation of the IOT system using Blynk application platform. This work proposed the low-cost interfacing module as alternative, and with the available commercial and open software, development could be made relatively easier. Data acquisition with LABVIEW DAQ card have been explored to many applications in remote monitoring. However due to cost issue, another alternative with lower development cost is preferable. Considering Arduino board as cost-effective solution for transforming the system capable to perform remote monitoring, thus elevates value to the use of LABVIEW to remain relevant. Taking example of remote experimentation framework at low-cost development for a temperature process control which can be accessed via student's mobile devices, the interfacing technique to be proposed in this work will benefit other researchers as a reference point to design their own data acquisition system in future.

Among all, the low-cost interfacing module have been studied in either for standalone or remote applications. Several previous works will present the current and latest interfaces used as data acquisition module to integrate between real applications to the LABVIEW. For a simple system, mostly adopted Arduino or Raspberry Pi boards Arduino is the highest among all.

2.3 Challenges and opportunities to real time execution in remote laboratory

Despite the fact that LABVIEW integration with Arduino demonstrated compatibility and minimizing device costs, the challenge in terms of integration process from physical laboratory to remote laboratory remains the most challenging so far, without denying that such project implementation is possible. However, interface integration between systems can be accomplished successfully with less effort if the issue of implementing remote laboratory into functional implementation is understood. The following are examples of such obstacles:

Article title	Application theme	Platform for interfacing hardware			
		Arduino	Raspberry-Pi	NI related-DAQ	PLC
Arduino and LabVIEW in educational remote monitoring applications	Education	•		•	
Remote monitoring and control of VFD fed three phase induction motor with PLC and LabVIEW software		•			•
Remote monitoring of BLDC motor using LabVIEW and zigbee	Healthcare	•			
An efficient IoT based biomedical health monitoring and diagnosing system using myRIO				•	
Development of a Medical Care Terminal for Efficient Monitoring of Bedridden Subjects		•			
Smart multi-level tool for remote patient monitoring based on a wireless sensor network and	Residential	•			
IoT application on establishing smart home		•		•	
Intellisense and remote centralized security monitoring system for the ventilation system in deep mining					•
Design and Real Time Implementation of SmartWater Management using LabVIEW and IoT	Agriculture			•	
An estimated method of visibility for a remote sensing system based on LabVIEW and Arduino	Energy	•			
Solar power remote monitoring and controlling using Arduino, LabVIEW and web browser		•			
Remote air quality monitoring system by using MyRIO-LabVIEW	Environment	•		•	

Table 2.
Summary of remote monitoring system using LABVIEW and interfacing hardware.

- Hardware configuration: The development the system, take into consideration the hardware that will be used, as this will impact the data flow process. Different setup configurations necessitate different initialization settings, necessitating the creation of a specific algorithm. To minimize errors, electronic components are chosen in such a way that they are configurable and easy to interface with.

- Complexity of data transfer: Transmission and storage become more expensive, while IoT adds complexity to the process of sending a continuous stream of data. These data will eventually affect the storage system.

- Internet connectivity: Since the internet connects the physical laboratory and the Blynk application, the laboratory has been enhanced to include the internet of things functionality. A connection to the internet allows the user to access the laboratory via smartphone or remote processing computer. However, the equipment's reliance on the internet has a downside when access is disrupted, or an internet interruption occurs.

As a result of executing the remote laboratory, the targeted programme can be further extended to support IR4.0 and be compatible with IOT. The use of open-source interfacing device provides advantages in terms of cost effectiveness, usability, and rapid prototyping to adjust the device design accustoms to applications and niche platforms built for particular use cases. This adaptation can increase the opportunities for educational institutions to form collaboration networks.

3. Methodology

To read, monitor, and control sensor data, LabVIEW employs a virtual instrument. MyRIO, DAQ, and NI-ELVIS are examples of known hardware interfaces. This hardware works on the same principle as the interfaces between the actual plant and the LABVIEW programming. Arduino and Raspberry-Pi are two of the most common data acquisition devices that support open-source programming by transforming functional interfaces into low-cost interface hardware. The use of LABVIEW with open-source hardware is gaining popularity due to increased practical implementation, especially in remote monitoring applications. The current study can be used as a pilot guide for developing a remote laboratory that is similar to an industrial-based temperature process. The proposed framework is intended to provide benefits in terms of practicality and cost-effectiveness. The built interface module in the proposed framework will provide access to the laboratory experimental setup, is illustrated in **Figure 4**.

Any experimental parameters or configuration input can be fed into the Arduino platform and transmitted to any user's mobile device connected to the laboratory network. The student can use this to remotely manipulate lab parameters and evaluate the outcomes without having to be physically present in the laboratory. This development's laboratory experiment involves data acquisition and PID control tuning of a modular-based LD-Didactic temperature equipment. A platform for reading and transmitting data and control parameters between the user and the remote laboratory setup is needed to design the module for this experiment (refer **Figures 5** and **6**). In addition, a user interface for displaying output that is accessible via mobile devices is required. Both temperature process modeling and PID controller tuning can be accomplished through algorithm development.

A mechanism for data collection necessitates an array sequence of data collection and transmission. When the thermocouple sensor reads the temperature of the oven, the input temperature transmits the data to the control unit of the processing computer, which is pre-installed with LABVIEW. Furthermore, prior initialization is required to establish LINX interconnectivity with a pre-programmed script of Arduino UNO using ATmega328P microcontroller operating at 16 MHz clock speed, a 32kB Flash memory, a 2kB SRAM, six analogue input, six I/O, one UARTs, one I2C, and one SPI. The gateway processes the received data and posts aggregated data with timestamp to the Blynk cloud. The temperature data are then stored in an 8-bit array and synchronized by sending it to the cloud. The current study can be used as a pilot template for establishing

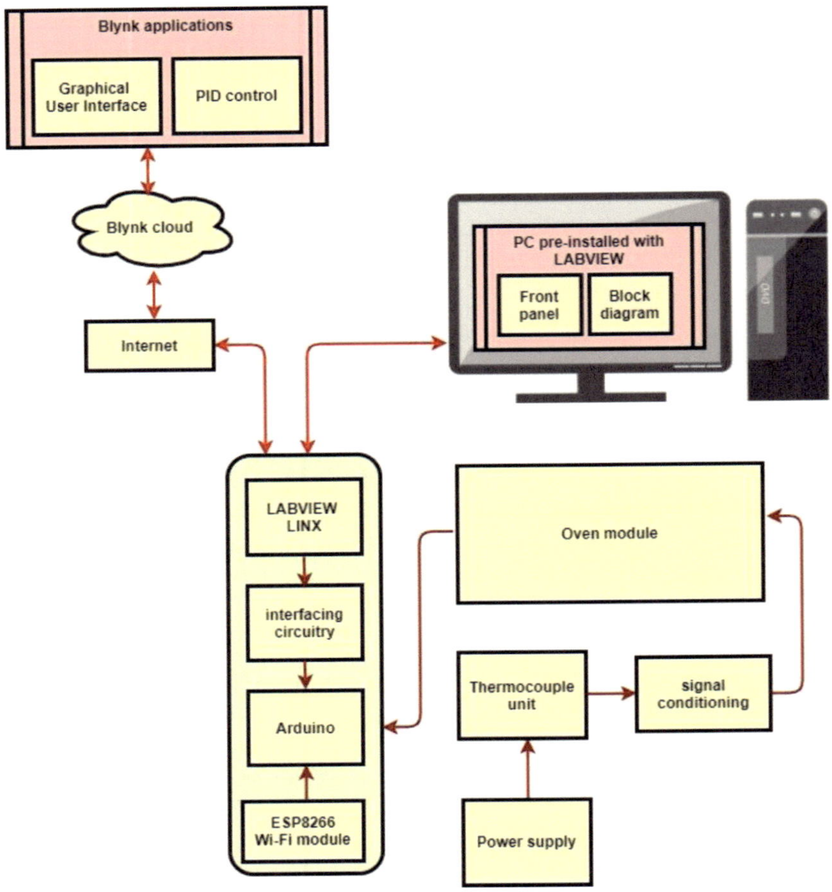

Figure 4.
Block diagram for modular based LD didactic temperature monitoring system and control with IOT.

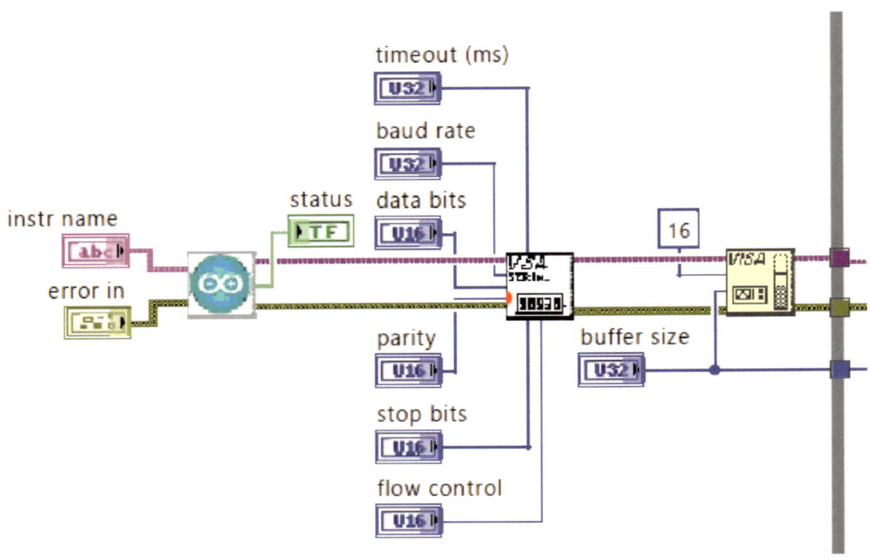

Figure 5.
LABVIEW's block diagram interfaces with Arduino module.

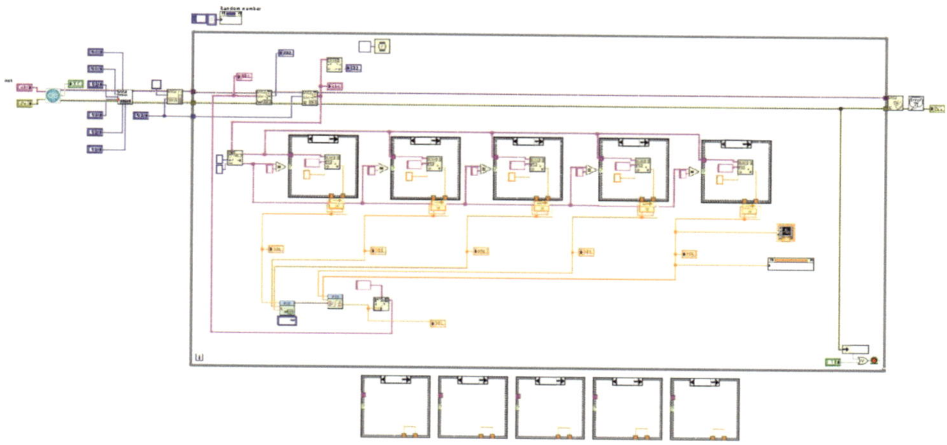

Figure 6.
Full LABVIEW's block diagram to perform interfacing with Arduino and PID control system.

a remote experiment between a temperature control plant and a mobile device to achieve remote laboratory application. The programming sequence is as follows:

Step 1: Initialize pin and load ESP8266 libraries in Arduino IDE.

Step 2: Setting up the network credentials: SSID and Passkey to establish IOT communications.

Step 3: Install NI VISA, VI Package manager

Step 4: Establish connection between Arduino board and LABVIEW. Then search for LabVIEW Interface for Arduino -- Firmware -- LIFA_Base. Conduct port setting and connect the board with LINX Firmware Wizard.

Step 5: Read temperature measurement and transfer to LINX Firmware Wizard.

Step 6: Transfer local data to cloud.

Step 7: Visualize measured data in Blynk application.

4. Blynk integration of IOT module with open-source hardware

Open-source platform for realizing the remote monitoring applications is increased nowadays. To achieve remote capabilities, there is a need to develop a low cost and user-friendly interfacing technique, which is suitable for communicating the physical experimental system and the student's mobile device in real time, realizable through IOT development. Among various options, Blynk application which is available with IOS and Android apps is capable to control the open-source hardware such as Arduino and Raspberry Pi. The interfaces in digital dashboard for IOT interfaces enable centralized data collection and analysis which is beneficial for laboratories applications. Even though there is existing IOT based data acquisition to perform remote monitoring function, however high cost of data acquisition device limits the actual implementation. Therefore, the proposed interfaces using Blynk application are the best economically available to display the temperature receive from this experiment conducted in laboratories.

Online display of data measurements is commonly dedicated for end user which accessible with phone or computer. In this case Blynk due to its simplicity of user interface programming by simply dragging-and-dropping widgets, network cloud can be configured conveniently, and access capability from both smartphone and laptop/computer. The best feature is to exploit the laboratory equipment while deploying the system away from the real experiment setup which can be monitored

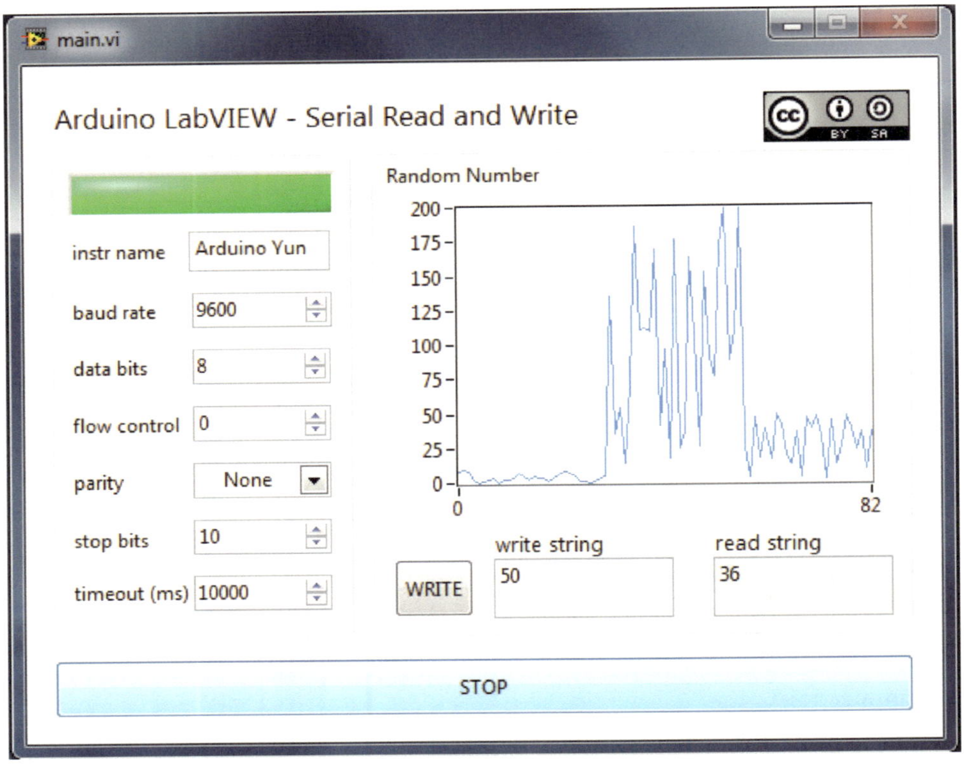

Figure 7.
VIs for Arduino-LABVIEW serial read and write.

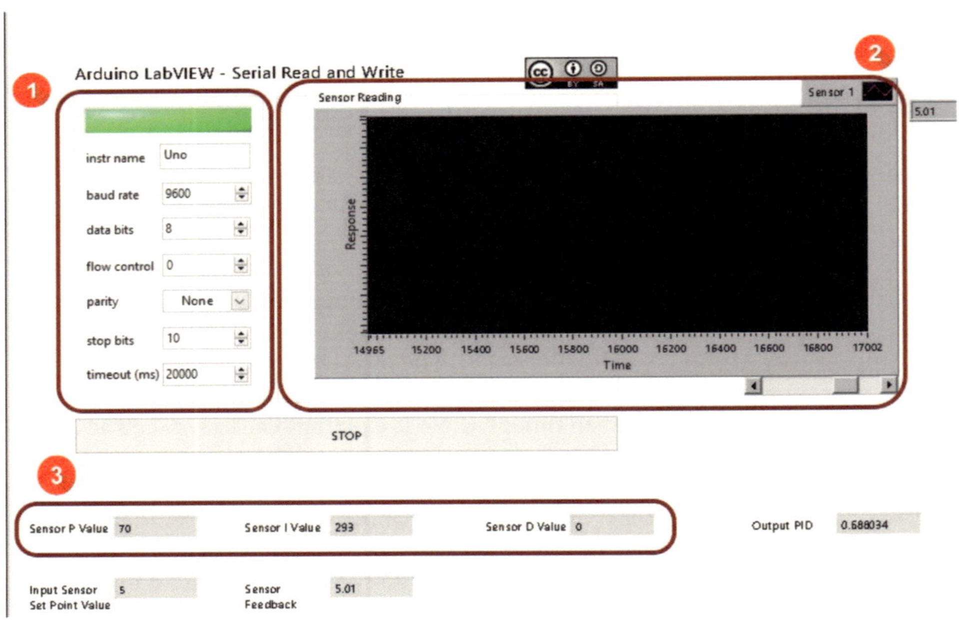

Figure 8.
VIs from the front panel of the PID control and monitoring system (1) Arduino-LABVIEW serial read and write (2) waveform chart (3) PID setting parameters.

and controlled from Blynk dashboard. The developed dashboard helps to provide better monitoring system with continuous monitoring and advanced data analysis which will be conducted later. Data acquisition process using Arduino-LabVIEW and IOT system need to be programmed in Arduino Integrated Development Environment (IDE) in terms of how to manage data sequence and arranging for data transferring process. While data retrieval from either the processing computer or plant to serve as data logging functions is controlled by Blynk, it involves the Blynk interfaces to be remotely monitored away from the experimental setup with internet connectivity. The collected data are made available to remote users in graphical form.

In order to build the data acquisition system with IOT interfaces, Blynk Arduino Library is required to install for establishing the firmware in the Arduino. To start with, BLYNK application requires to be configured and registered using an email. An authentication code with the SSID and the password of the network station is created uniquely for each of developed, and thus need to write these details as part of programming code. As shown in **Figure 7**, the design created for remote monitoring application to apply in temperature experiment study contains four slider widgets to represent the variables used for tuning process and a SuperChart widget to display real-time graph (see **Figure 8**).

5. Real time PID tuning control for LD didactic temperature plant

In addition to the existing control engineering laboratories, integrating an industry-relevant process with a remote lab has increased the understanding of temperature process modeling and tuning scheme. Obtaining accurate temperature control is crucial when simulate the PID tuning parameters namely K (proportional gain), Ti (integral time) and Td (derivative time). **Figure 10** demonstrates the experimental setup of the temperature measurement system. The plant contains several modular types of thermocouple unit, signal conditioning unit, meter, and processing unit to preinstall with LABVIEW. Real-time tests are shown in **Figure 11**, have revealed that the acquired measurement of plant temperature with several PID setting can be observed from the

Figure 9.
Blynk interface with widgets configured for remote monitoring.

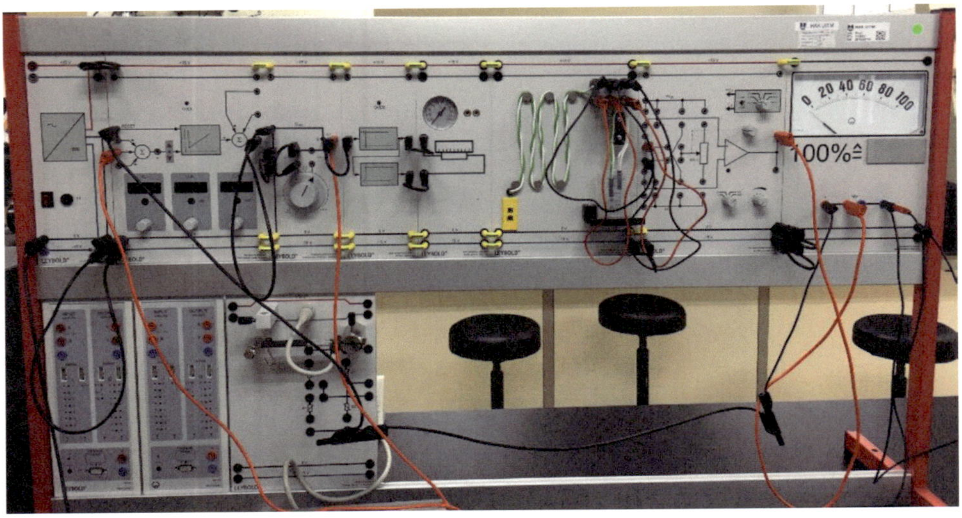

Figure 10.
Experimental setup.

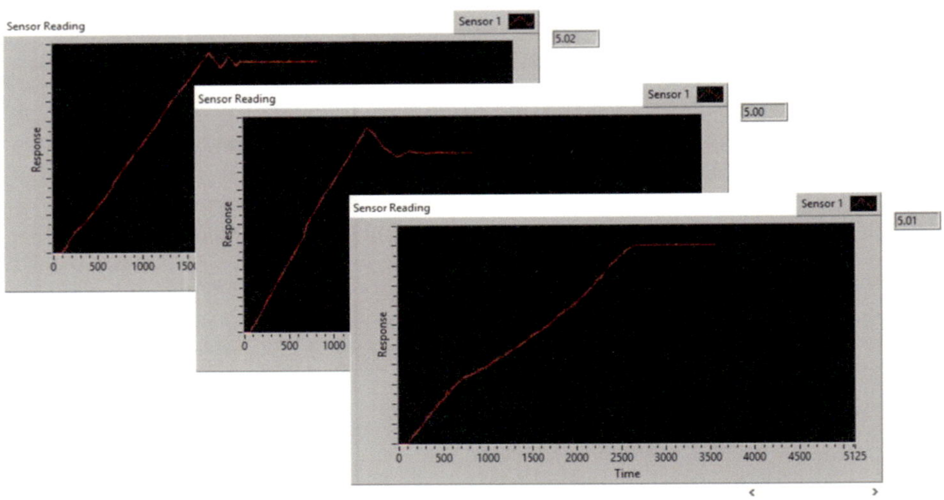

Figure 11.
Simulations results on PID tuning.

produced graph. The result based on the graph which represented from the PID tuning demonstrate the working of Blynk application to transfer and receive temperature data is acceptable and reliable. Moreover, the alert system was designed to alert when our hardware interface disconnects from the Blynk (**Figure 9**). Thus, IOT interface has made possible with the presence of low-cost module. Additionally, this study will also provide learning experience in creating a framework to digitize an existing process for monitoring purposes.

6. Conclusions

This chapter describes data collection and device integration into IOT applications for remote monitoring. Coordination of data transfer and aggregate technique is used to ensure the effectiveness of the interfacing technique for realizing the remote laboratory. Using an open-source feature with LINX and Arduino microcontroller board as the gateway of data communicating between plant and mobile device, the advantage and challenge were presented, benefiting the end user in terms of more realistic hands-on and a better understanding of the method. Following up on recent progress, an improved version of the remote monitoring for temperature sensing will be published, along with new models and applications.

Acknowledgements

I would like to thank Universiti Teknologi MARA for providing support throughout the research works.

Conflict of interest

The authors declare no conflict of interest.

Author details

Muhammad Asraf Hairuddin[1], Nur Dalila Khirul Ashar[1*], Amar Faiz Zainal Abidin[2] and Nooritawati Md Tahir[3]

1 College of Engineering, Universiti Teknologi MARA, Cawangan Johor, Kampus Pasir Gudang, Malaysia

2 Faculty of Electrical and Electronics Engineering Technology, Universiti Teknikal Malaysia Melaka, Melaka, Malaysia

3 College of Engineering, Universiti Teknologi MARA, Shah Alam Selangor, Malaysia

*Address all correspondence to: nurdalila306@uitm.edu.my

References

[1] M. Hernández-de-Menéndez, A. V. Guevara, and R. Morales-Menendez, "Virtual reality laboratories: a review of experiences," Int. J. Interact. Des. Manuf., vol. 13, no. 3, pp. 947-966, 2019.

[2] M.-H. Zhang, C.-Y. Su, Y. Li, and Y.-Y. Li, "Factors affecting Chinese university students' intention to continue using virtual and remote labs," Australas. J. Educ. Technol., vol. 36, no. 2, pp. 169-185, 2020.

[3] C. Arguedas-Matarrita *et al.*, "Remote experimentation in the teaching of physics in Costa Rica: First steps," in *2019 5th Experiment International Conference (exp. at'19)*, 2019, pp. 208-212.

[4] A. Benhamouda, B. Benmounah, N. Baira, and S. Kahmous, "Design and Implementation of a Low-Cost and Modular Remote Lab Framework: Application to Electronic Sensors," in *International Conference on Interactive Collaborative Learning*, 2017, pp. 657-664.

[5] A. Cardoso, V. Sousa, M. T. Restivo, and P. Gil, "Demonstration of a remote lab based on a vibrating beam apparatus," in *2016 13th International Conference on Remote Engineering and Virtual Instrumentation (REV)*, 2016, pp. 357-358.

[6] A. Cardoso, J. Leitão, P. Gil, A. S. Marques, and N. E. Simões, "Using IPython to Demonstrate the Usage of Remote Labs in Engineering Courses–A Case Study Using a Remote Rain Gauge," in *International Conference on Remote Engineering and Virtual Instrumentation*, 2018, pp. 714-720.

[7] D. Galán *et al.*, "Safe Experimentation in Optical Levitation of Charged Droplets Using Remote Labs.," *J. Vis. Exp. JoVE*, no. 143, 2019.

[8] P. Abreu, J. S. Valiente, L. De La Torre, and M. T. Restivo, "Remote experiments with pneumatic circuit using a double rod cylinder," in *2019 5th Experiment International Conference (exp. at'19)*, 2019, pp. 410-414.

[9] I. Angulo *et al.*, "RoboBlock: A remote lab for robotics and visual programming," in *2017 4th Experiment@ International Conference (exp. at'17)*, 2017, pp. 109-110.

Chapter 4

LabView and Connections with Third-Party Hardware

Giuseppe Porzio

Abstract

Data acquisition is a function that plays a fundamental role in the automatic supervision and system control, it combine the system (software and hardware) to the process to be controlled (real world). The field of application starts from research to automation, from industry to home automation, in practice everything that in some way must be performed without human supervision. Data acquisition systems are mainly used to measure physical phenomena such as: temperature, voltage, current, distance and pressure, shock and vibration, and displacement, RPM, angle and discrete events, weight. In order to measure it we need a DAQ, Data AcQuisition System, in this chapter we propose to use a cheap open source hardware: Arduino.

Keywords: Arduino, cheapest hardware, wiring code, LabView code, producer/consumer, SCADA system

1. Introduction

Data acquisition is a function that has a role of fundamental importance in the functions of automatic supervision and control because it relates the system (software and hardware architecture) with the process to be controlled (real world). The field of application ranges from research to automation, from industry to home automation, basically everything that in some way must be performed without human supervision.

Data acquisition systems are primarily used to measure physical phenomena such as: temperature, voltage, current, strain and pressure, shock and vibration, distance and displacement, RPM, angle and discrete events, and weight.

When the engineer is interested in controlling a physical process (light intensity, sound analysis, mass measure, position check, velocity, PID control, etc.) his first problem is to acquire the right information coming from one or more sensors, in some cases we talk about sensor strings or distributed sensors.

The goal is to acquire data that are consistent over time and that correctly describe the shaping of the physical process. All this allows both the correct processing of data and a fast action on the control system through its actuators (motors, LEDs, speakers, etc.).

"Data acquisition" means data exchange in both directions: from the process to the system and vice versa. In all control systems the "heart" of the process is the data acquisition that plays a main role but at the same time it must be accompanied by a simple and intuitive user interface, the HMI-Human Machine Interface. Data acquisition systems are generally referred to by the acronym DAQ (Data AcQuisition).

IntechOpen

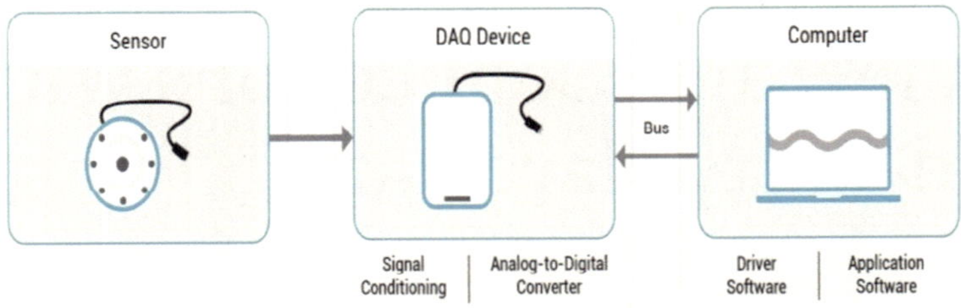

Figure 1.
(Acquisition chain) [1].

Figure 1 shows the electronic chain to acquire an analog signal. The sensor is the device sensitive to the physical feature, the analog-to-digital conversion system, and the computer on which the SW architecture for managing the information is developed. Both feedback and actuators are missing in this figure as they are not the subject of this chapter.

This chapter is designed to be a guide for beginners, programming amateurs and students who wish to approach the world of automation with LabView using low-cost third-party DAQs such as Arduino.

Arduino is a "machine" capable of working in Stand Alone, it can perform simple industrial control tasks.

In a SCADA (Supervisory Control And Data Acquisition) system there is a Master and many Slaves. The Master device carries out the configuration, supervision and control of the slaves. The slave, a local device very close to the process, is equipped with a processor and a system of ports to interface with the sensors and actuators. In this chapter we will write some code to have LabView in the role of Master and Arduino in the role of Slave.

2. Sensor, filtering and multiplexer

We speak about Data Acquisition process, DAQ, when we refer to the process of making measurements of physical phenomena with a PC (tablet, smartphone, workstation, etc). The signals, to be processed, are converted from the analog domain to the digital domain. Only after the digital acquisition we can process the data acquired (recording, visualization, analysis). For this purpose, an A/D (Analog to Digital) subsystem is used to convert the signal.

We report, below, some theoretical hints of the components visible in **Figure 2**.

At the sensor output, the electronic chain includes a "signal conditioning circuit", a multiplexer, the sampling circuit and finally the A/D converter.

The measurement of a physical phenomenon, such as temperature, sound level, vibration of motion oscillatory, or wind speed, begins with a sensor. A sensor is a device that converts the physical phenomenon into a measurable electrical signal.

For example, an elevator gets to the floor through the installation of positioning sensors; a washing machine is equipped with a sensor that measures the rpm of the motor or the water level in the drum; a twilight light; a TV remote control. The classic mercury thermometer is also a type of sensor that is used to measure temperature. In this case, however, the measure is expressed directly on a graduated scale readable by man and not by the machine: we speak in this case of **human readable** type sensor.

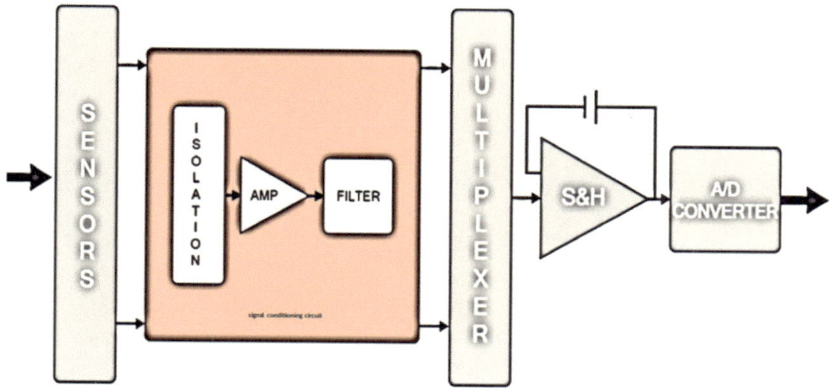

Figure 2.
Detail of the complete acquisition scheme.

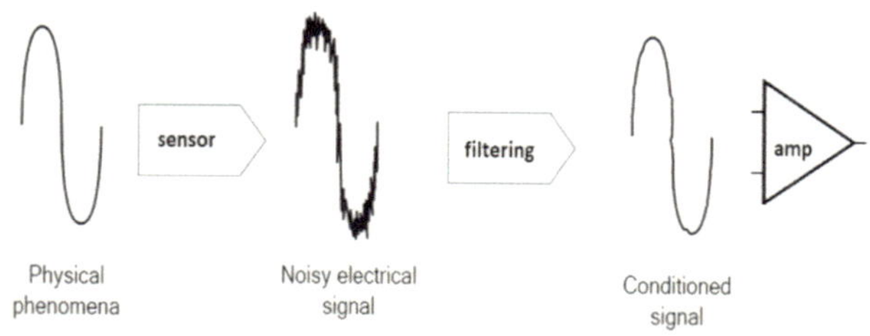

Figure 3.
Signal conditioning, filtering and amplification.

The sensors can produce several kind of electrical outputs such as voltage, current, resistance, or other electrical characteristics modulated from physical phenomenon. When the signal coming from the sensor or from the transmission line is noisy or the ground reference is not at 0 volts (as it should) is preferable to use an isolation system.

In "signal conditioning circuit" we propose a section with electrical isolation that allows the separation of the signal from other electrical sources. This aspect is also essential for the measurement of signals with very small amplitude in which external electrical potentials can affect the quality of the signal considerably, providing incorrect results.

2.1 Signal conditioner

The **signal conditioner** circuits are designed to process the analog signal from the sensors and prepare it to be digitally sampled. The conditioning circuit must linearize the sensor output, eliminate electrical interference that adds to the signal (so-called "noise", as shown in **Figure 3**), and amplify the small signal (mV, μV) to a nominal level, to be easily digitized.

2.2 Multiplexing

Multiplexing, on the other hand, is that part of the circuit that allows us to expand the inputs of our DAQ, thus using a single conversion line on multiple input channels **Figure 4**.

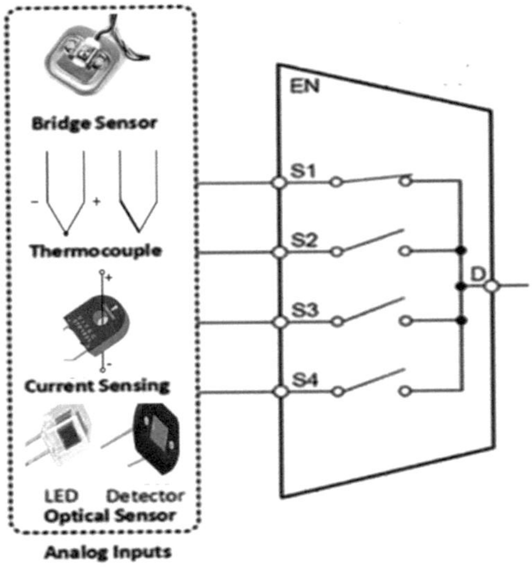

Figure 4.
Example of a 4-input multiplexer.

The multiplexer (commonly called MUX) is a selector of data lines (analog or digital) able to select different input signals: once selected the channel, the corresponding signal is collected and sent on the output line. There are some particularly performing and expensive devices that do not use the MUX but they have a complete acquisition chain for each input.

3. Sampling and coding

3.1 Sample and hold

The S&H system is the circuit part that performs the sampling of the signal (sampling phase). Sampling, in signal theory, is a technique that consists in converting a continuous signal in time into a discrete signal, evaluating its amplitude at regular time intervals. Therefore, considering that the physical quantity attributed to the physical phenomenon varies continuously over time without any interruption, it is necessary to decide with which time interval to interrogate the sensor in order to have meaningful data for our measurement.

From the definition of the sampling interval (T_c) for the scan we derive the sampling rate:

$$f_c = \frac{1}{T_c} \tag{1}$$

The effect of the circuit in **Figure 5** is to store the analog value taken at a given time (sample phase) and keep it constant for as long as it takes the converter to perform the conversion (hold phase).

But how fast should the sampling rate be? Clearly it depends on the phenomenon we are observing. See two examples below:

1. We want to monitor the temperature of a room to stabilize it at a value of T_{set} ± error. Considering the inertia of the room and the radiators it makes sense to acquire the temperature every second i.e. f_c = 1 Hz.

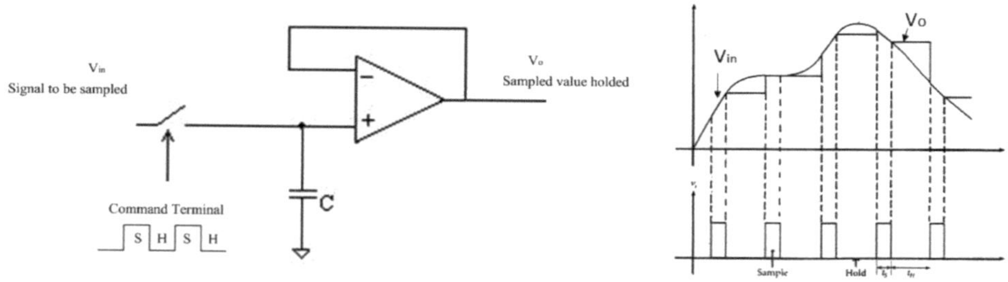

Figure 5.
S&H circuit and example.

Question:

2. How high has to be the sampling rate if we would like to create automatic braking for anti-collision car system? Assume that max velocity, for small/medium sized car, is 180 km/h.

Answer:

$$180\,km/h \overset{:3,6}{\Longrightarrow} 50\,m/s \tag{2}$$

Let us assume that the control system reacts in such a time that the car still travels at maximum for 10 cm (**response time**).

So, if we make some calculations, the time between one reading and the next one of the vision sensor must be less 2 ms. These involves

$$f_c > 500Hz \tag{3}$$

The proposed cases are at the antipodes: while in the first one we do not have any criticality, in the second one there is a big responsibility due to the need to manage the stop of the car before the impact.

In the real world, according to the mathematician J. Fourier, an analogue signal can be represented by linear combination of sinusoidal functions (called **harmonics**).

The first harmonic, called fundamental, has the same frequency as the input signal, while the following harmonics will have a frequency multiple of the fundamental.

The transition from the analogue to the digital domain, therefore discrete, leads us to acquire one of these harmonics, of course the first one, therefore a suitable sampling frequency will be the key to a good acquisition of the analogue signal, preserving its main characteristic, its frequency.

3.2 Sampling theorem (or Nyquist-Shannon theorem)

In order to have a correct sampling (without loss of information) we must to choose a correct frequency of sample rate. Supposing that the frequency signal (first harmonic) is f_{max} then Nyquist-Shannon Theorem says that the minimum sampling frequency f_s, that preserve the frequency information of a original signal, should be double of the fmax.

$$f_s \geq 2f_{\max} \tag{4}$$

If, in addition to the frequency, we would like to storage also the shape of the signal we need:

$$f_s \geq 5 f_{max} \quad (5)$$

The sampling rate is normally expressed in Sample Rate and the unit of measure is number of samples per second [#S/s].

To understand better how the theorem works in the **Figure 6** we report a sequence of acquiring with several sampling rate. The software used is developed for university student's lectures [2].

Figure 6 shows how the sampling frequency acts. We start with a 440 Hz source signal (resonance frequency of a conventional tuning fork) which is visible in the first waveform graph of the sequence. In the following sequences the following sampling frequencies were used: 440 Hz, 600 Hz, 880 Hz and 2200 Hz.

In the first and second cases the fs is not adequate, in fact we have an under-sampling. In the third case we have a result that preserves the frequency of the input signal. Finally, in the last case, we have reconstructed quite faithfully the profile of the original signal.

3.3 Coding

The last sequence in the DAQ chain (**Figure 2**) consists of the operations performed by the A/D converter: quantization and encoding. First we need to introduce the concept of signal dynamics. The dynamics of the signal indicates the maximum excursion of the signal and, therefore, also the maximum and minimum values it can reach, the range of V_{in} (also defined as the Full Scale value):

$$V_{FS} = V_{max} - V_{min} \quad (6)$$

(we have assumed a voltage signal)

The input signal, being continuous in time, can by definition take on an infinity of values. As well as the sampler has discretized the signal in time (X axis) we now need another circuit which discretizes the values of the physical quantity which represents the information (Y axis). So the technique is to approximate the value acquired in the sampling phase to a discrete value. The number of discrete values available for these approximations is given by a very simple calculation. If we choose *n* bit to make a digital conversion then the number of discrete value is 2^n.

At this point we have to define the unit of quantization that we call quantum or quantization step, that is the smallest approximation interval that we use to compare the sampled signal to discretize it.

$$Q[V] = \frac{V_{FS}}{2^n} \quad (7)$$

Q is called quantization step. It is possible to assert, at this point, that a higher bit number and a smaller V_{FS} interval implies the greater number of intervals available. This means that the size of the interval will tend to be an extremely small value with increasingly accurate measure.

The simplest coding (commonly used for unipolar signals, i.e. always positive ones), natural binary code (straight binary), consists in making each quantization interval correspond to a progressive binary number, starting from 0 (corresponding to the lowest level) up to 2^n-1.

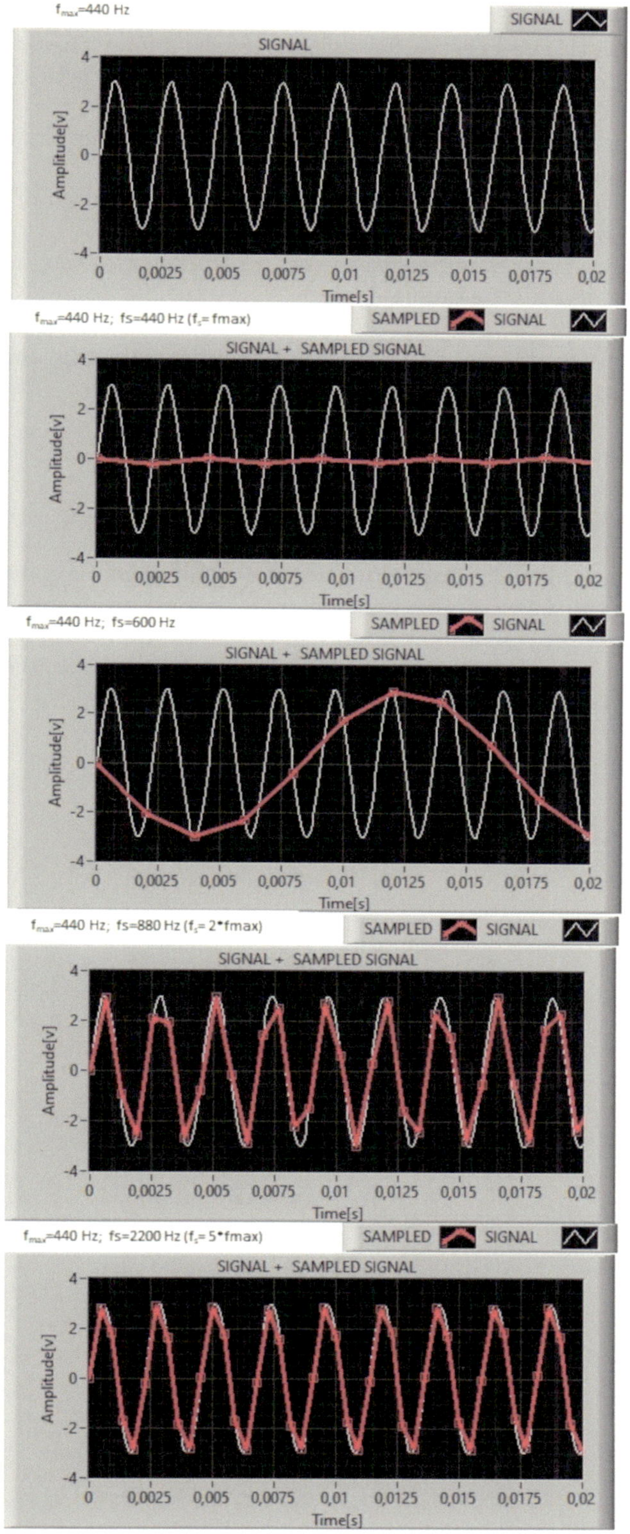

Figure 6.
Signal sampled with different f_s.

In **Figure 7** we show what we have said, on the X-axis we put the intervals between V_{max} and V_{min} and beside them the bit combinations. The first level consists of all bit to zero, so the word 000...00 corresponds to V_{min} while the last level is given by the word with all ones 111....11 i.e. V_{max}.

A different number of resolution bits clearly produces different quantization ranges, some data is shown in **Figure 8**.

Clearly the measurement of Q is affected by error and corresponds precisely to Q/2 and is defined as quantization error.

Recapitulate, in order to perform a correct measurement through a DAQ system, the following points must be satisfied:

1. Prefer a sensor with a linear response and that the maximum and minimum values are compatible with the dynamics of the DAQ;

2. Choose an appropriate sampling rate;

3. Choose an appropriate resolution;

The premises made so far are useful to better understand the code written for the Master unit and the slave unit. In this chapter we propose an cheap and open source prototyping board for which we will write some code to transform it into a DAQ. The proposed board is Arduino UNO rev.3. In the next paragraph, the Arduino technology will be presented [3].

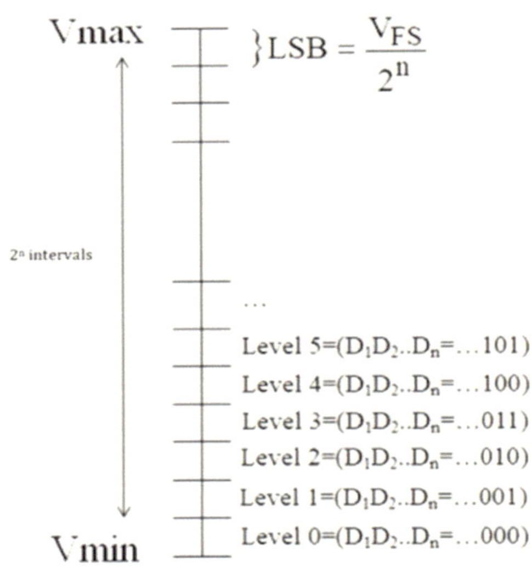

Figure 7.
Quantization and coding.

	12 bit	16 bit	18 bit	24 bit
#level	4.096	65.536	262.144	16.777.216
input range ± 10 V	4,88 mV	305 μV	76,4 μV	1,192 μV

Figure 8.
Resolution example.

DOI: http://dx.doi.org/10.5772/intechopen.96056

4. Arduino UNO rev. 3

Arduino Uno (**Figure 9**) is a microcontroller board (Italian open source project) based on the ATmega328P (resolution @10 bit; input range 0÷5 V). It has 14 digital input/output pins (of which 6 can be used as PWM outputs), 6 analog inputs, a 16 MHz like internal clock (sample rate = ~10 kS/s), a USB high speed connection, a power jack 9 Volt input, an ICSP header, reset button and several states LED like Tx/Rx serial communication.

It contains all interfaces needed to support the microcontroller and its functionality; You can use prototype board with your Uno without worrying about doing something wrong, worst case you can replace chip with a new one and start over again. The Uno board is the first USB Arduino boards, today are available several models of it: with wifi o ethernet, compact or large model, wearable, etc.

Wiring is an open-source programming framework for microcontrollers C/C++ based.

The developer, under conditions of classical use, writes code for Arduino in order to have a "machine" that works in Stand Alone, in **Figure 10** is shown his working scheme, the code runs on Arduino, through the code reads the sensors and produces actions on the physical world. In the next paragraph will be discussed the code to transform Arduino from Master to Slave.

The new role of Arduino will be to be used in LabView environment as a real data acquisition system (**Figure 11**).

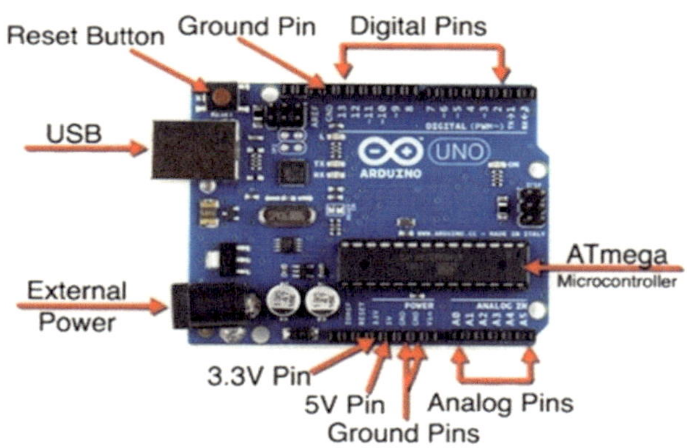

Figure 9.
Arduino UNO rev.3.

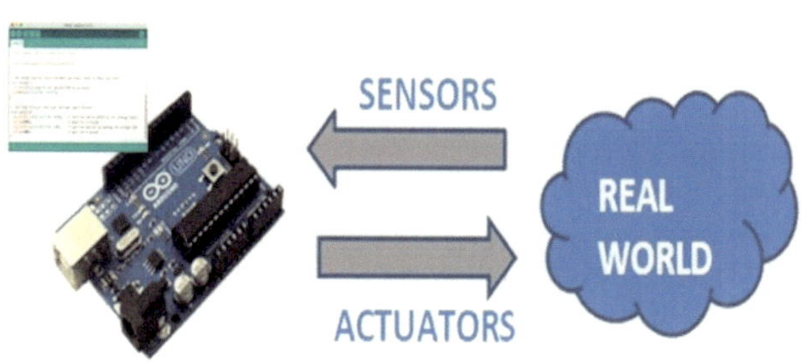

Figure 10.
Arduino-stand alone mode.

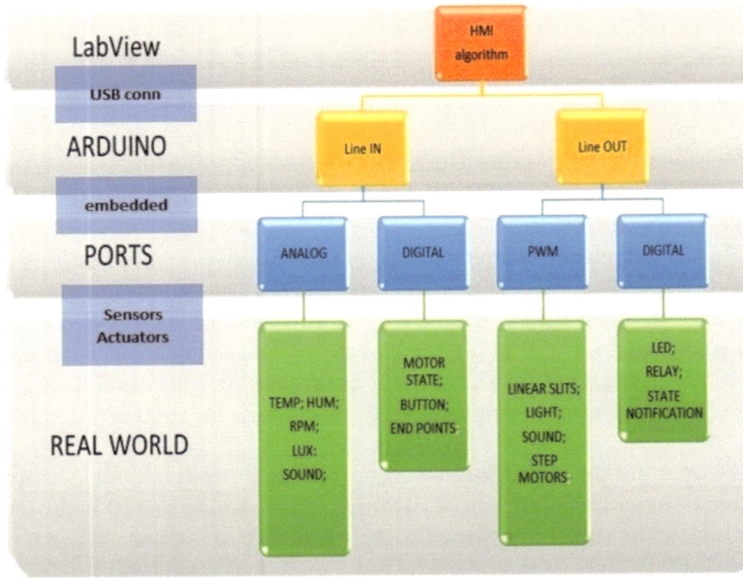

Figure 11.
Control hierarchy with LabView-Arduino.

4.1 From master to slave

Among of programmer "sketch" is the name that Arduino's programmer uses for a program. It's of code written in like C, compiled and, then, uploaded on the board. After it is possible to run on an Arduino board the code. There are two distinct functions available in Arduino sketch: *setup()* and *loop()*.

The *setup()* is called once time only at beginning when the sketch goes in run. It's a correct place to make setup tasks like setting pin modes or initializing libraries.

The *loop()* function is a infinite loop and is heart of most sketches. You need to include your algorithm and functions in it.

Normally in the setup() section there is the sequence of instructions to configure all the Arduino peripherals and features that will be used in the project such as: Analog input, PWM, i2c. In loop(), instead, is written all the control algorithm that will be characterized by an infinite loop.

In this paragraph we propose the development of a code from a different perspective, Arduino will be used as a DAQ system. So inside the setup() there will be a pre-cycle in which the Arduino waits for the USB connection to LabView and waits for the ASCII character sequence to configure the Arduino ports as desired.

The ASCII code, we call op-code from now, to send for configuration are printable characters, so you can always test the Arduino code from any serial terminal or using the serial monitor of the IDE.

For example, to configure the Analog Input channel zero (A0) just send the code "a". Arduino will remain in the setup() section until the master sends the character "z" on the serial which will end the setup cycle to execute the code in the loop().

The code proposes a scenario in which analog inputs A0÷A5, DIO pin2 and pin4 and a PWM channel on pn3 are configurable. Clearly it is possible to extend the "offer" by adding other input or output lines. The complete management of a sensor through Arduino libraries could also be included.

Regarding the sampling time Ts it is possible to define through the ASCII codes A,B,C,D a time delay equal respectively to 100 msec, 10 msec, 1 msec, 500 μsec. If it is omitted the acquisition time is 1000 msec.

```
/* SerialEvent occurs whenever a new data comes in the hardware serial RX. This routine
is run between each time loop() runs, so using delay inside loop can delay response.
Multiple bytes of data may be available. */

void serialEvent() {

  // get the new messagio from master

  String input_from_LV = Serial.readString();

  if(input_from_LV == "R\n"){

   Serial.println("Reset, you have to wait...");

   Reset();

  }

  if(input_from_LV == "D0_ON\n"){

   Serial.println("DIGITAL ZERO_ON");

   digitalWrite(2, HIGH);

  }

  if(input_from_LV == "D0_OFF\n"){

   Serial.println("DIGITAL ZERO_OFF");

   digitalWrite(2, LOW);

  }

  if(input_from_LV == "D1_ON\n"){

   Serial.println("DIGITAL UNO_ON");

   digitalWrite(4, HIGH);

  }

  if(input_from_LV == "D1_OFF\n"){

   Serial.println("DIGITAL UNO_OFF");

   digitalWrite(4, LOW);

  }

  if((input_from_LV.substring(0,3)) == "PWM"){

   val=input_from_LV.substring(3,6).toFloat();

   analogWrite(3, val);

  }

  Serial.println(input_from_LV);

}
```

```
// blinking function- Called during
the channel configuration

void blinking() {

 for (int i=0; i<2;i++){

   // turn the LED on (HIGH is the
voltage level)

   digitalWrite(LED_BUILTIN,
HIGH);

    // wait for 150 msecond

   delay(150);

  // turn the LED off by making
the voltage LOW

   digitalWrite(LED_BUILTIN,
LOW);

   delay(150);

}

}
```

Figure 12.
Code- SerialEvent() and blinking().

The code developed in the loop() section collects data from the previously configured input line ports, maps them to the following format #A0#A1#A2#A3#A4#A5$D0$D1 and sends the message continuously to the USB port. The message will contain as many strings as there are lines configured. In the syntax #Ai (i = 0...5) the value of Ai corresponds to the decimal decoding of the combination of the 10 bits, so there will be 2^n combinations. At value 0 will correspond 0 (zero) Volt and at value 1023 will correspond 5 Volt.

WE suggest to the reader to test own system velocity before to set 500 µsec of sample rate. Usually, for my experience, it is very rare to follow with a LabView (not real time) Loop code that velocity. In case the system is not fast enough, one way of not losing data could be the following: change the Arduino's code to collect the msg (measured value) in a vector of 100 elements and send it to LabView each 50 msec. You can choose different size of vector but you have avoid to saturate the Arduino memory.

The op-code (operation code) we have written does not belong to any standard communication protocol. We have invented a sequence of simple ASCIII strings to be sent over serial. So the Master will have at his disposal a set of instructions, which can be extended by the reader, to change the status of a digital output: D0_ON\n, D0_OFF\n, D1_ON\n, D1_OFF\n.

In order to avoid a slowdown loop() for sensors reading, due at continuous polling on the receipt of messages from the Master, an event-driven solution has been considered.

The reception on the serial line of a request from the Master is triggered by the event generated by the chip that manages the USB communication. When a byte arrives on RX an event is generated and triggered by a software procedure. When this occurs the Master message will be read (**Figure 12**).

In the end we can send a message to set a Analog output by pin3 in PWM mode.

The **Pulse Width Modulation** [4], or PWM, is a powerful technique to control analogic circuits (applied to a load) using a digital signal. It is a type of digital modulation, in particular we speech of pulse width modulation which allows to obtain a variable average voltage depending on the ratio between the duration of the high pulse and the entire period (duty cycle).

In electronics it is used to change the voltage, and therefore the power, on a generic load. For example, to change the speed of a direct current electric motor, to vary the brightness of light bulbs, especially LEDs. A useful duty cycle of 0% indicates a pulse of zero duration, in practice no signal (V_{out} = 0 volts), while a value of 100% indicates that the pulse ends when the next one begins ($V_{out} = V_{cc}$). To use

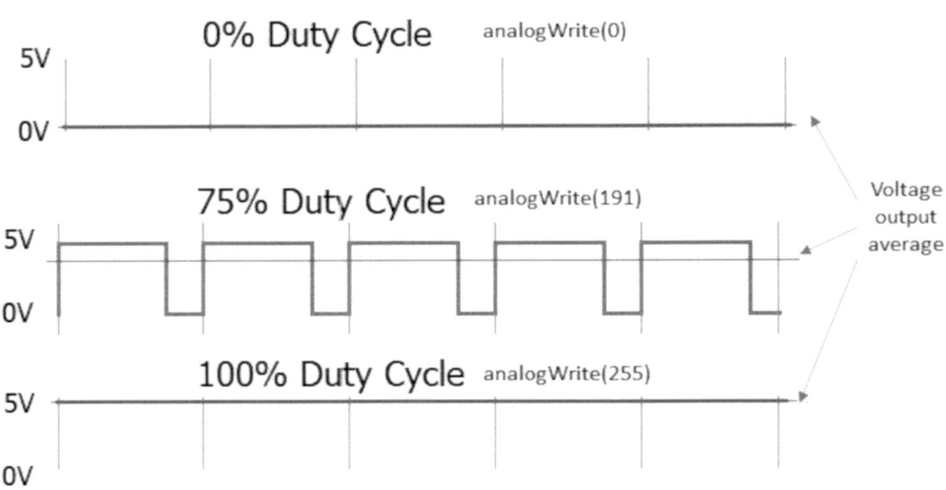

Figure 13.
PWM example.

this technique with Arduino is very simple, with the analogWrite (PIN, VALUE) function it is possible to modulate the work cycle. The PIN corresponds at PWM pins and VALUE is scale from 0 to 255. For example analogWrite (pin, 255) corresponds to a 100% duty cycle and analogWrite (191) is a 75% duty cycle (**Figure 13**).

4.2 Arduino code

In the following boxes (**Figure 15**) we'll show some code that you can use to create a communication Master–Slave from LabView and Arduino [5].

In **Figure 14** it is possible to understand the functionally of declarations reading the comments.

In **Figure 15** is possible to verify the **setup()** code, inside it there are the comments to understand it. The code in **Figure 15** has been conceived to be very static and the expansion is very simple for the novice programmer. The serial speed has been set at $2 * 10^6$ bit/s in order to have the maximum communication speed between Arduino and the Master.

The code written in the "WHILE LOOP" could be redesigned to treat the Arduino channels dynamically. We want to say that configuration strings (like **pinMode(4, OUTPUT)**) can be sent directly from the MASTER unit.

```
// This Code is written frm G. Porzio for chapter of
// "LabVIEW - History, Applications, Current Trends and Perspectives",
// ISBN 978-1-83968-841-6.
//----------------------------------------------------------------------
// Variable Declarations
int opCode= 0;          // incoming serial byte
int Ts=1000;            // Sampling time [msec]; inizialize at 1 second
String output;          // msg to send at serial port with analog value
// each bit corresponds a sensor, the value at 1 means the sensor is present the LSB bit = Analog Sensor in A0
byte Sensors = 0;
// each bit corresponds a Digital line, the value at 1 means that the line is configurated as input.
byte digital_line_OUT=0;
// each bit corresponds a Digital line, the value at 1 means that the line is configurated as output.
byte digital_line_IN=0;
// each bit corresponds a PWD line, the value at 1 means that the line is.
byte Analog_write_PWM_LB=0;
String input_from_LV=""; // String contains the op-code
int val;
void(* Reset)(void) = 0; // is a code for software RESET
//----------------------------------------------------------------------
```

Figure 14.
CODE-variable declaration.

```
void setup() {
  // initialize digital pin LED_BUILTIN as an output
  // We use it to blink when Arduino recevices the string configuration from LaBView
  pinMode(LED_BUILTIN, OUTPUT);
  digitalWrite(LED_BUILTIN, LOW);   // we turn it OFF

  // start serial port at 2Mbit/sec and wait for port to open:
  Serial.begin(2000000);
  Serial.flush();
  while (!Serial) {
    ; // wait for serial port to connect.
  }
  // we send a msg to LabView
  Serial.println("connection estabilized");

  delay(500); //  wait 500 msecond
  /* the following code is used to configure, in this example,
  the analog input A0-A5, the digial I/O (pin 2 and pin 4),
 and PWM line (pin 3) of course it is possible to expand
 it as want as you wan
  We use in the follow "SWICTH CASE" statement The ASCII code:
  1)a,b,c,d,e,f to configure A0-A5;
  2)g,h to Configure the pin 2 and pin4 like digital in;
  3)i,l to Configure the pin 2 and pin4 like digital out;
  4)m to configure the pin 3 like digital out and
 we'll use in PWM mode;
  5)A,B,C,D to set the Ts (Sample rate) at 100 msec,
 10msec, 1 msec, 500 usec;
  6) z to finish the setup().
   */
   while (opCode!=122){ // exit with statement "z"
    // get incoming byte:
     opCode = Serial.read(); //reads the Bytes sequence on USB
     switch (opCode){
       case 97 :  // "a"
       Serial.println("channel A configurated");
       bitSet(Sensors,0);
       blinking();
       break;
       case 98:  // "b"
       Serial.println("channel B configurated");
       bitSet(Sensors,1);
       blinking();
       break;
       case 99:  // "c"
       Serial.println("channel C configurated");
       bitSet(Sensors,2);
       blinking();
       break;
       case 100: // "d"
       Serial.println("channel D configurated");
       bitSet(Sensors,3);
       blinking();
       break;
       case 101: // "e"
       Serial.println("channel E configurated");
       bitSet(Sensors,4);
       blinking();
       break;
       case 102: // "f"
       Serial.println("channel F configurated");
       bitSet(Sensors,5);
       blinking();
       break;
```

```
case 103: // "g"
       Serial.println("Digital 2 is input");
       pinMode(2, INPUT_PULLUP);
       bitSet(digital_line_IN,0);
       blinking();
       break;
       case 104: // "h"
       Serial.println("Digital 4 is input");
       pinMode(4, INPUT_PULLUP);
       bitSet(digital_line_IN,1);
       blinking();
       break;
       case 105: // "i"
       Serial.println("Digital 2 is output");
       pinMode(2, OUTPUT);
       bitSet(digital_line_OUT,0);
       blinking();
       break;
       case 106: // "j"
       Serial.println("Digital 4 is output");
       pinMode(4, OUTPUT);
       bitSet(digital_line_OUT,1);
       blinking();
       break;
       case 107: // "k"
       Serial.println("Digital 3 in PWD mode");
       pinMode(3, OUTPUT);
       bitSet(Analog_write_PWM_LB,0);
       blinking();
       break;
       case 65:  // "A"
       Serial.println("set Ts at 100 msec");
       Ts=100;
       blinking();
       break;
       case 66:  // "B"
       Serial.println("set Ts at 10 msec");
       Ts=10;
       blinking();
       break;
       case 67:  // "C"
       Serial.println("set Ts at 1 msec");
       Ts=1;
       blinking();
       break;
       case 68:  // "D"
       Serial.println("set Ts at 500 ucsec");
       Ts=0,5;
       blinking();
       break;
       case -1:  // null
       case 10:  // "\n"
       case 122: // "z"
       break;
       default:
       Serial.println("channel error");
       break;
      }
```

Figure 15.
Code-slave mode setup().

```
Serial.flush();
  while (input_from_LV.substring(0,5)) != "EXIT"){
   String input_from_LV = Serial.readString();
  // Master send, for example, pinMode 5 OUTPUT, we have to parse the phrase
   if (input_from_LV.substring(0,7)) == "pinMode"){
  pinMode(input_from_LV.substring(8,9), input_from_LV.substring(10,16))
  // put your code here

   blinking();
   }
```

Figure 16.
Example code for dynamic configuration.

DOI: http://dx.doi.org/10.5772/intechopen.96056

```
/* In the loop() code we use the bitRead to Know if the channel
 * is configurated in the setup() statement
 */
void loop() {
output= String ("");
// Serial.println(Sensors);
 if (bitRead(Sensors, 0)==1){
    output+="#";
    output+= String (analogRead(A0));
 }
 if (bitRead(Sensors, 1)==1){
    output+="#";
    output+= String (analogRead(A1));
 }
  if (bitRead(Sensors, 2)==1){
    output+="#";
    output+= String (analogRead(A2));
 }
 if (bitRead(Sensors, 3)==1){
    output+="#";
    output+= String (analogRead(A3));
 }
 if (bitRead(Sensors, 4)==1){
    output+="#";
    output+= String (analogRead(A4));
 }
 if (bitRead(Sensors, 5)==1){
    output+="#";
    output+= String (analogRead(A5));
 }
 if (bitRead(digital_line_IN, 0)==1){
    output+="&";
    output+= String (digitalRead(2));
 }
 if (bitRead(digital_line_IN, 1)==1){
    output+="&";
    output+= String (digitalRead(4));
 }
Serial.println(output);

delay(Ts);          The "delay()" behaviour is
                    like the sample rate

}
```

Figure 17.
Loop() code.

Figure 16 shows a small piece of code that is a good starting point for completing the dynamic channel configuration.

At this point we show the code about the blinking procedure and the Serial events procedure, respectively both in **Figure 12**.

In the end we report the loop() code, **Figure 17**. The code is very simple, the final message is made-up by concatenating the message in each "if" statement.

5. LabView architecture

In this section we show you the architecture that we use to run LabView code in Mater mode. We have chosen the *Producer/Consumer Architecture* [6].

The Producer/Consumer design pattern (**Figure 18**) is based on the Master/Slave pattern, and is geared towards enhanced data sharing between multiple loops running at different rates. The Producer/Consumer pattern is commonly used when acquiring multiple sets of data to be processed in order. Suppose you want to write an application that accepts data while processing them in the order they were received. Because queuing up (producing) this data is much faster than the actual processing (consuming), the Producer/Consumer design pattern is best suited for this application. In our project we can set a high sample rate (up to f_s = 10 kHz) so in this can we can occur in a data loss case. With Producer/Consumer is sure that we are implementing a data lossless LabView architecture. But we have considerated also an architecture *Event-Driven* to catch the write instance from LabView vs. Arduino only if asked from the operator.

In **Figure 19** we show the front panel developed in LabView [7].

5.1 Front panel

On the left side of front panel are present a several controls to configure the DAQ (Arduino in Slave mode) in according with previous paragraphs. Instead on the right side we found a control to set the PWM value (analog output) and the digital output state. We use Waveform chart like oscilloscope to view the six signal.

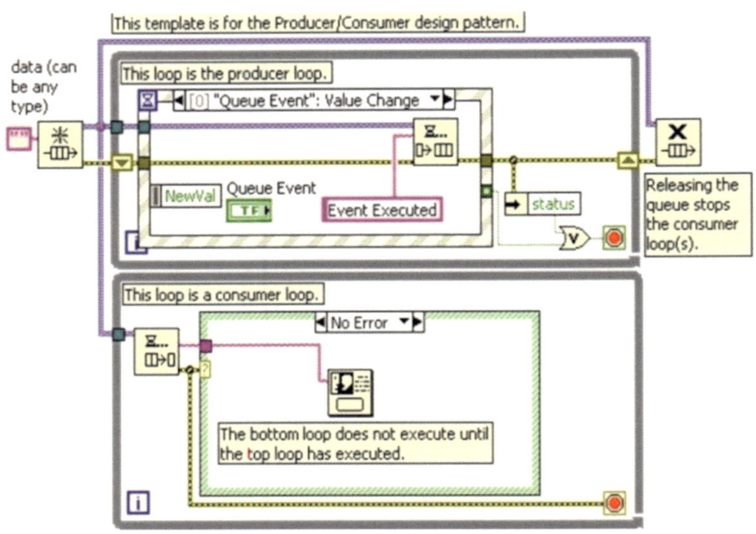

Figure 18.
Event structure in producer/consumer design pattern.

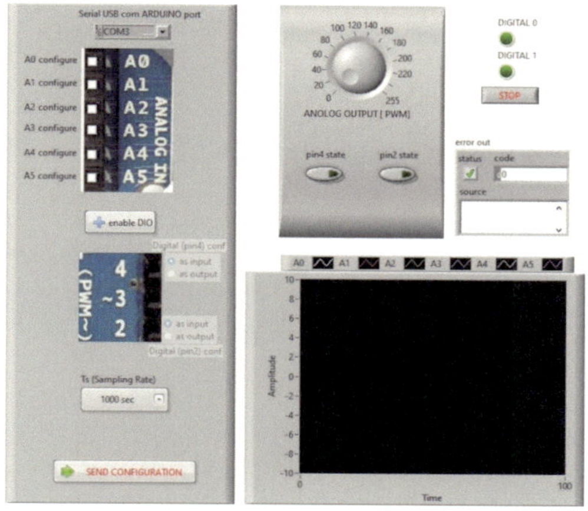

Figure 19.
LabView front panel.

After defined the configuration you have to send a message at the serial VISA communication by the pressing of "SEND CONFIGURATION" button.

After that the cycle Producer/Consumer starts and the sensor reading is shown on the Waveform Chart.

5.2 Block diagram

Inside the LabView Code (block diagram) there are three nodes. The first one composed by "While Loop" (**Figure 20a**) that waiting for user's hardware configuration. In this Loop we create a Boolean array with all hardware instance, at the end of the configuration the user pushes the button and send the array to subvi "open and configure.vi" (second node). It makes a rights sequence of op-code, open the Serial Port communication (in this case **com3**) and send it at Arduino (**Figure 20b**). During this phase you can observe the blinking LED on Arduino board, this means that the configuration message has correctly reached Arduino and it is processing the op-code.

From **Figure 20b** is possible to verify that the serial port velocity is 2Mbps.

In this way the communication between Master and Slave does not make interference with acquiring. In fact one character, in ASCII encoding (1 byte), from Arduino to LabView is sent in 4 μsec. If we would configure all analog inputs (6) and all digital inputs (14) the maximum number of characters would be = 6 prefixes (#) + 6*4 (digits of value among 0÷1023) + 14 prefixes (&) + 14 digital states = 58 bytes.

Maximum time to transmit the entire message is 58 Byte * 4 μsec = 232 μsec. This time is half of the minimum sampling time set in the code, that is 500 μsec. You could also reach 100 μsec of sampling rate that corresponds to 10 kHz of sampling frequency, in this case you have to merge the bits of the digital input, so it is possible to save 26 bytes but it is not enough. We have to modify the syntax of sending analog input values to reach at least 80 μsec of transmission time. This modification to the Arduino code we leave to the reader as an exercise.

In last one node, **Figure 21**, we can see the Producer Loop and the Consumer Loop. Both are connected by the queue, in queue process we read the Arduino's

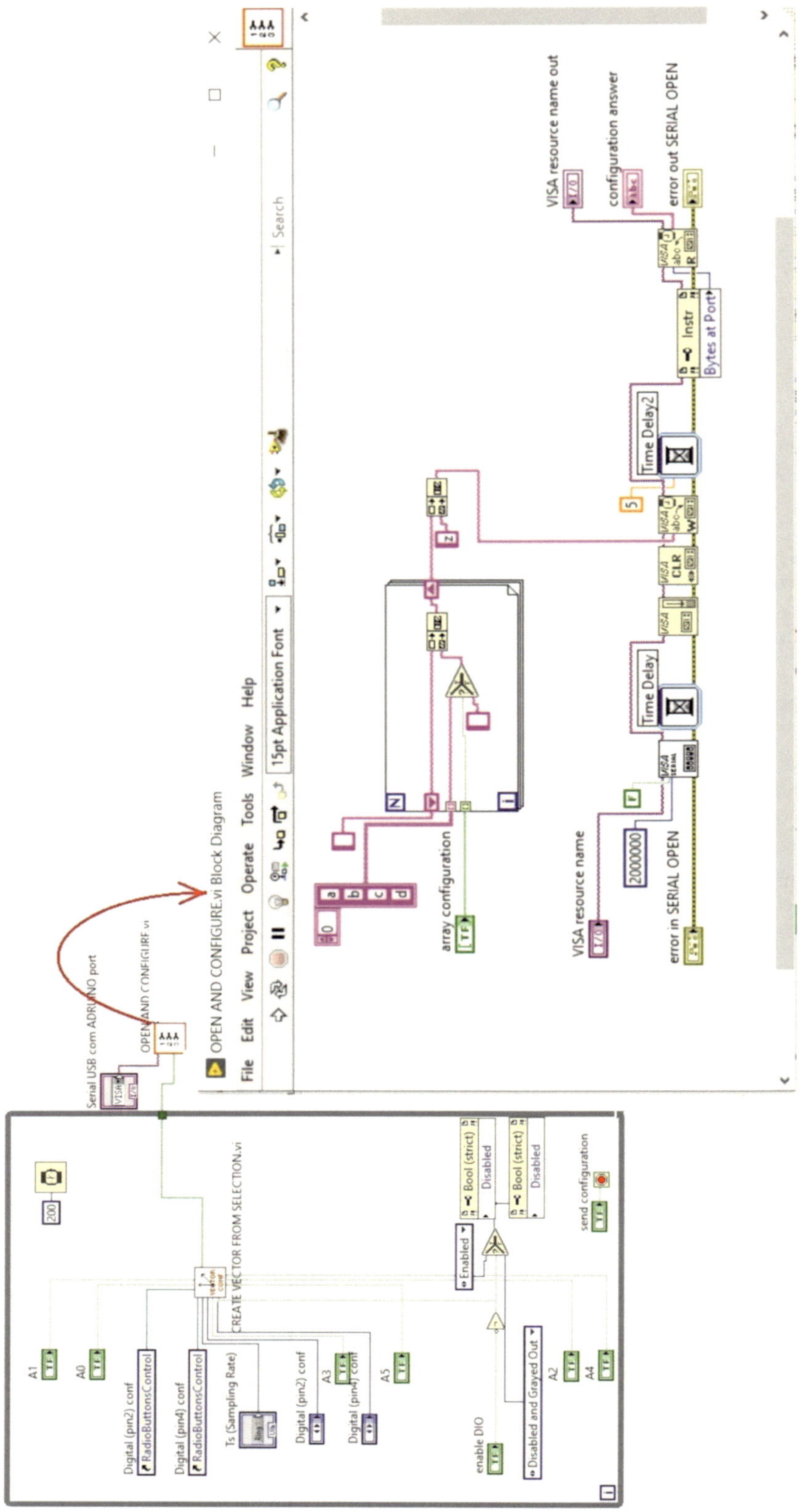

Figure 20.
First node and second node in block diagram (a and b).

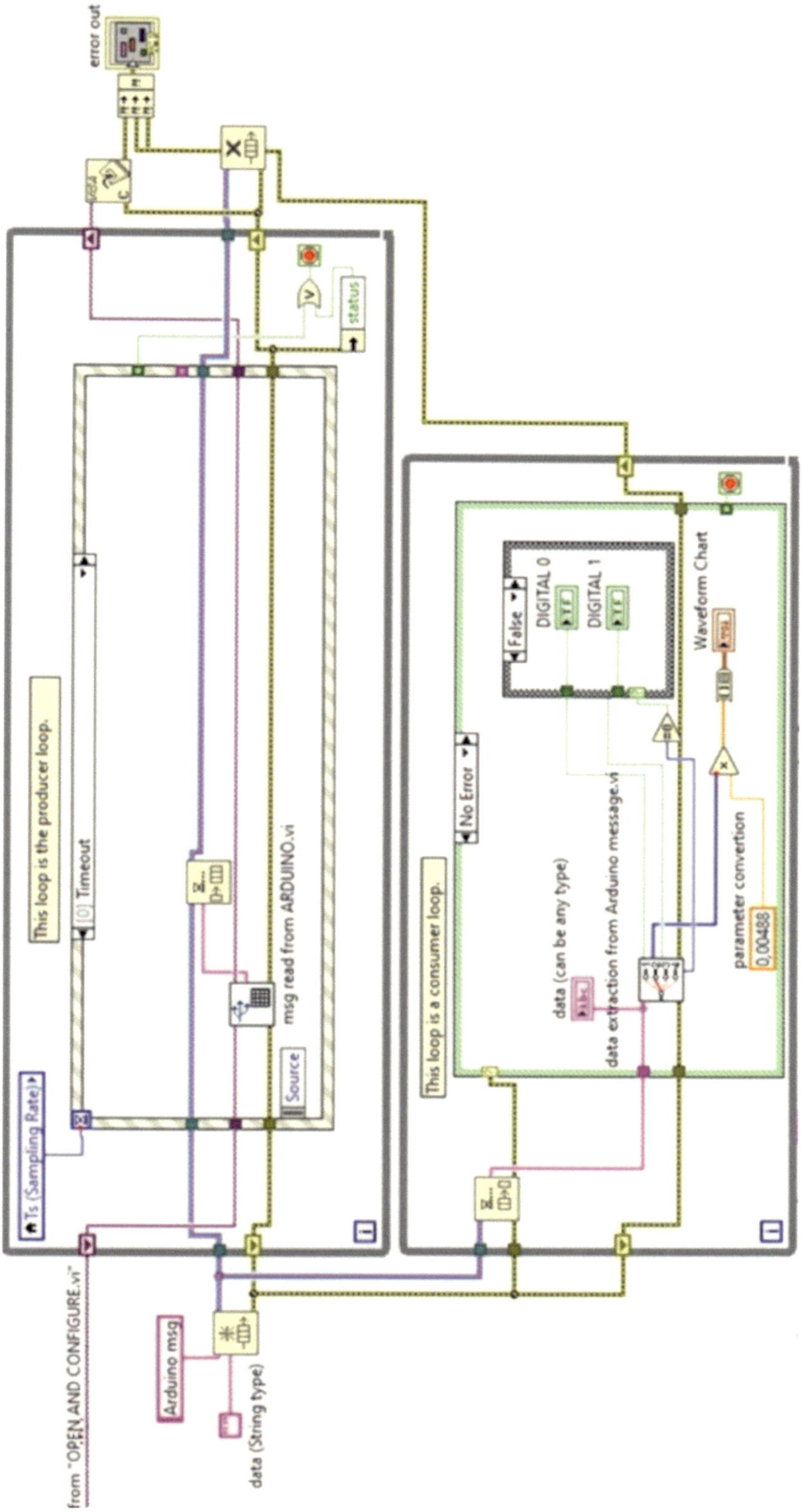

Figure 21.
Producer/consumer event-driven LabView CODE.

message at maximum frequency and by consumer loop we process the data. The Event-Driven statement is configurated with the following **events.**

5.3 Timeout

The timeout terminal of "Event Structure" is connected, of course, at local variable"Ts (Sampling Rate)" in according with sample rate configurated in Arduino in node 1. In this "case" we read with "msg read from ARDUINO.vi" the Arduino's message from serial (**Figure 22**). It is very simple code. The data are available on serial port (hardware) and the code read it using a **Bytes at Port** function.

5.4 Analog output [PWM]

in this case we send a message to Arduino by serial port, remember that Arduino reads the message with a SerialEvent() function. Here we make a message with a

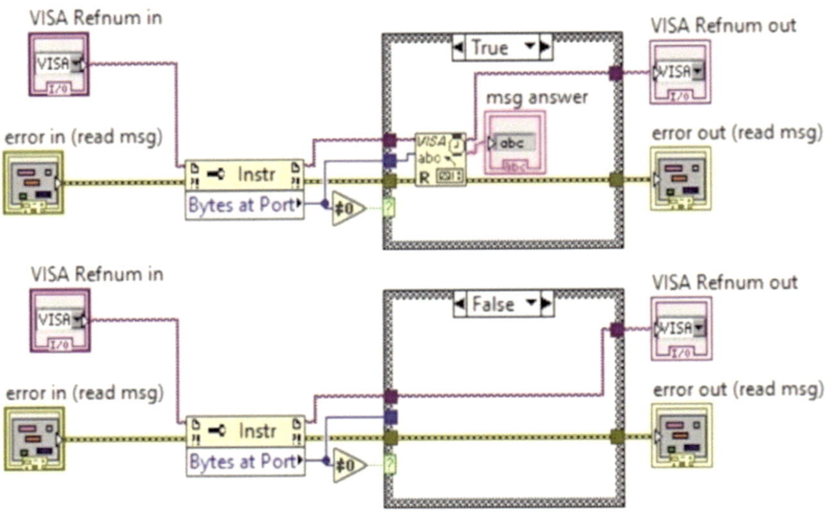

Figure 22.
Block diagram of msg read from ARDUINO.Vi.

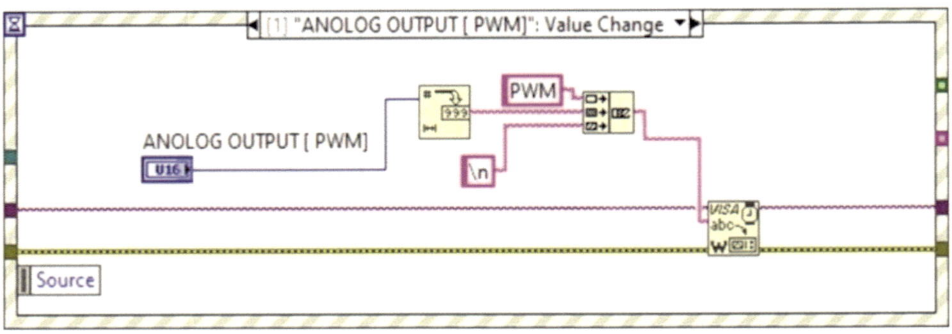

Figure 23.
Analog output [PWM] code.

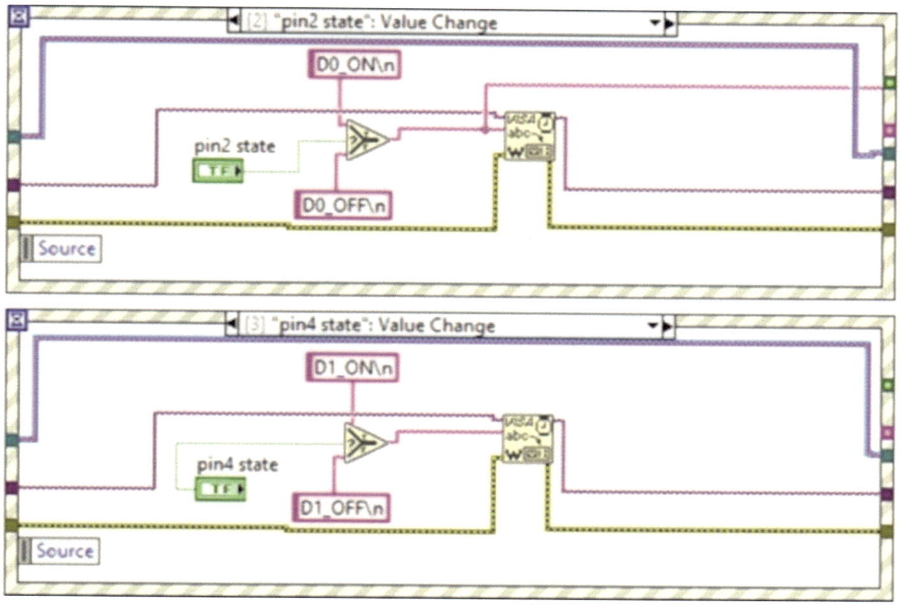

Figure 24.
Pin2 & pin4 event.

word PWM followed with "ANALOG OUTPUT [PWM]" control knob converted in ASCII code (**Figure 23**).

5.5 Pin2 state and Pin4 state

In this "case" (**Figure 24**) we build the message to send Arduino by serial port to change the digital pin state, remember that Arduino reads the message with a SerialEvent() function **Figure 12**.

Now we go back at **Figure 21** where we have to talk about the Consumer Loop. Through the enqueue function we read the data from the head of the queue with the FIFO method (first in first out). If we have not error the data read are processed with the subvi "data extraction from Arduino message.vi".

In **Figure 25** there is the screen code. The code scan the message, check if present special ID char (#) or (&) and collect the data by **indexing** it on the loop edge. With **Conditional indexing** we choose where collect the data: Analog Array or Digital Array.

The subvi "data extraction from Arduino message.vi" returns the status of the digital inputs and the numerical values of the analogue inputs, if configured.

To convert the integer values reads from analog ports we need to perform a simple conversion. According to what we have studied in the previous paragraphs having a 10 bit ADC and a dynamic of 5 volts we obtain:

$$\frac{5[V]}{2^{10}} = 0{,}00488[V] \tag{8}$$

At this point, in the consumer loop, before displaying the analogue signals on the Waveform chart we multiply the output by the value 0.00488.

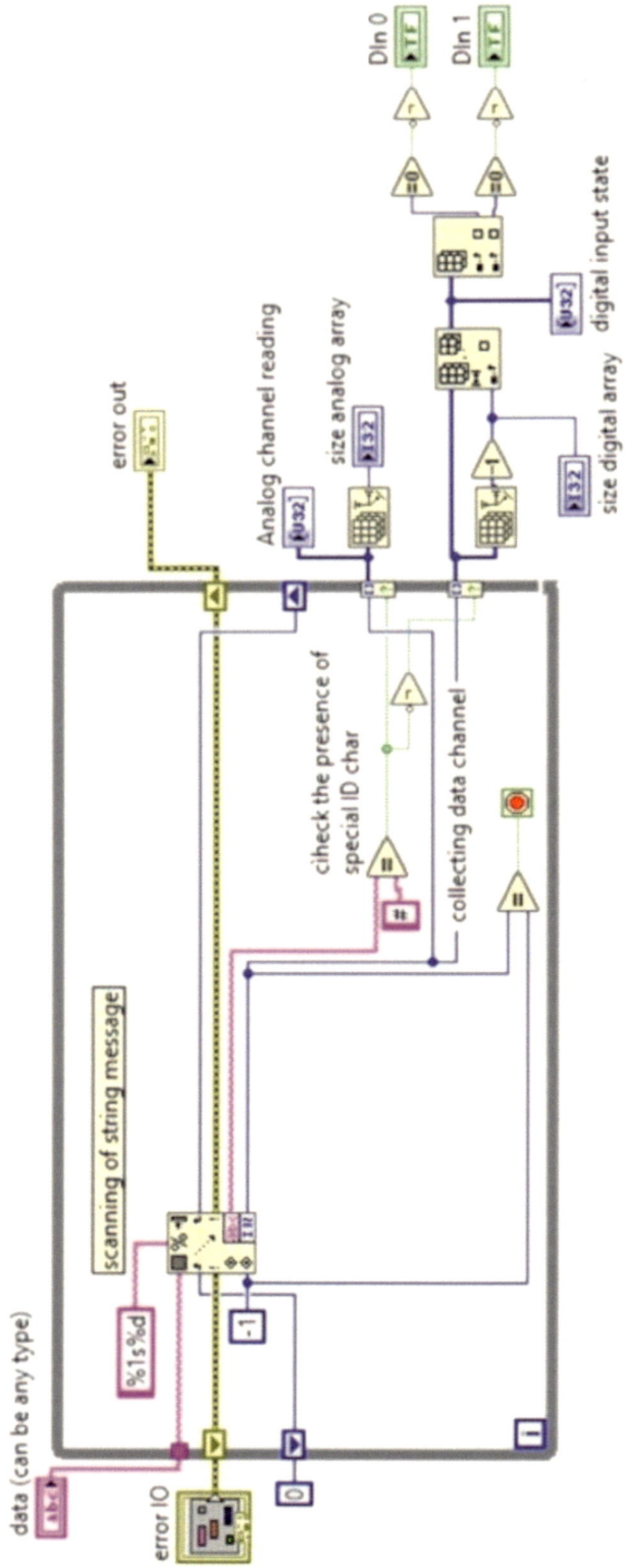

Figure 25.
Data extraction from Arduino message.vi.

DOI: http://dx.doi.org/10.5772/intechopen.96056

6. Conclusions

In this chapter we have seen one of the many ways of how LabView can be used with third parties hardware. The idea is to have an inexpensive tool not for industrial use but for High School applications where it is possible with a few euros to set up a laboratory for the analysis of an RC/RLC circuit, voltage divider, diode/transistor characterization. With a cheap sensors, connected at Analogue inputs, you can prepare laboratory experiments such as the pendulum oscillation, spring characterization, measurements of angles in uniform angular motion, etc.

In the end you could organize LabView CORE I and CORE II training courses in e-learning where the DAQ board is very cheap and easily purchased on the web from the students.

Conflict of interest

The authors declare no conflict of interest.

Thanks

Dedicated to My wife and my daughters for encouraging and supporting me.

I would like to thank, my friend, the Director of the Department of Mathematics and Physics at my University, Prof. Lucio Gialanella, for supporting my initiative and for his precious advice.

Author details

Giuseppe Porzio
University of Study Of Campania "Luigi Vanvitelli", Caserta, Italy

*Address all correspondence to: giuseppe.porzio@unicampania.it

References

[1] LabVIEW™ Core I PN 326292A-01

[2] Giuseppe Porzio is the author of LabView SW for didactic experience

[3] Alan GS. Introduction to Arduino: a piece of cake. ISBN: 1463698348

[4] Jian S. Dynamics and Control of Switched Electronic Systems Pulse-Width Modulation. pp. 25-61. ASIN: B00A9YGCWC

[5] Available from: https://github.com/gporziog/LabView-and-connections-with-third-party-hardware/tree/master/Ardunio_Like_DAQ

[6] LabVIEW™ Core II PN 326293A-01

[7] Available from: https://github.com/gporziog/LabView-and-connections-with-third-party-hardware/tree/master/LabView%20Master%20CODE

Chapter 5

LabVIEW and Open Embedded System

Trinh Quang Duc

Abstract

In this chapter, discussions about the applications based on LabVIEW with a typical open embedded system such as Arduino. Instead of the applications with use of National Instrument hardware, an user can build an low-cost microcontroller-based system which user interface and procedure designed in LabVIEW. Recently, Arduino-based applications are one of interested developments, the embedded system is designed with the low-cost microcontroller of ATMEGA2560 and easy to program with IDE (Interface Development Environment). However, the difficulties of Arduino-based applications are user interface design while LabVIEW is an excellent utilities for the panel designs. The limitation of some LabVIEW applications is the requirements from combination with high performance NI hardwares. Under interactions between LabVIEW and IDE library designed by Arduino, the low-cost system can be easily built for experiments or prototype designs.

Keywords: LabVIEW Embedded Systems, Biosignal Acquisition, LIFA toolbox, LINX toolbox, LabVIEW Arduino Interfaces

1. Introduction

Embedded system requires the interaction between hardware and software installed in PC (Personal Computer). Before LabVIEW, a software designed for hardware devices seems to be hard work job because of the communication protocols are need to be programmed in both hardware device and PC. For instant applications which aim to measurement data acquisition for analysis algorithm development or process control as well as controller design, the hard work in programming field may restrict the destination job. To solve the problem, National Instrument introduced LabVIEW as a graphical programming tools with a series of compatible hardware devices [1]. This tool has been widely chosen by the non-professional user community. The NI hardware communicate with PC through some communication protocol such as CAN (Controller Area Network), GPIO (General Purpose Input/Output), SPI (Serial Peripheral Interface), PROFIBUS (Process Field Bus), PROFINET (Process Field Network) and LIN (Local Interconnect Network), are officially equipped with software called driver licensed by National Instrument. The independent hardware developments for LabVIEW compatibility, hence, are not easy.

Recently, some multiple purpose low-cost open-source hardware modules based-on micro-controller were developed by commercial companies such as Arduino [2], NodeMCU [3]. This device connects with PC using USB port and transceivers data through a simple embedded program set on the flash memory of the micro-controller. With data acquisition and control applications which the discrete sampling as well as response frequency not over than 100 kHz, this device will be the considered options for experimental arrangement. Since the first time released in 2005 [4], Arduino is used to develop many applications in automation systems. However, until 2013, a proposal for LabVIEW interface toolkit as LIFA (LabVIEW Interface for Arduino Toolkit) was announced [5] and introduced [6]. Another interface named as LINX designed by Digilent [7]. So far, these toolkit is widely known for the users using Arduino and Raspberry Pi in LabVIEW application for embedded system.

The LabVIEW VI is programmed to communicate with hardware devices though a package called NI-VISA [8]. This package provides some interface with communication protocol such as PXI, GPIB, VXI, Serial, USB. The LabVIEW compatible hardware devices has to be set their communication procedures in the flash or ROM memory to handshake with the VI program, then, the data can transfer from the devices to PC and the command can receive from the PC to the devices. The main problems of any embedded system is how to build the communication procedure set on the device to match with the communication procedures installed in LabVIEW designed by National Instrument. This is solved by some independent projects such as LIFA and LINX which the communication procedure is designed.

2. LINX and LIFA and the embedded model

LIFA and LINX are the open-source but their approaches are different. LIFA provides a base succeed procedures installed in the file named as LabVIEWInterface.h, the users, based-on this procedure can program other procedures in the hardware devices to control their functions. To use LIFA, beside the graphical programming in LabVIEW, the procedure programming for the device firmware is required. On the other hands, LINX considers the LabVIEW graphical programming rather than the firmware programming, the compatible firmware are designed the loaded to the correspondent hardware devices respectively, therefore, with a simple functions, the embedded system with LINX is simpler than use of LIFA. Because of that, LINX is better error handling but does not flexible to develop functions like LIFA. For some unique embedded system, LIFA is preferred than LINX.

Figure 1 show diagram of typical embedded system using LINX and LIFA. To use LINX, users should install the package into LabVIEW as a function of the LabVIEW and then choose a compatible firmware to load into the hardware flash

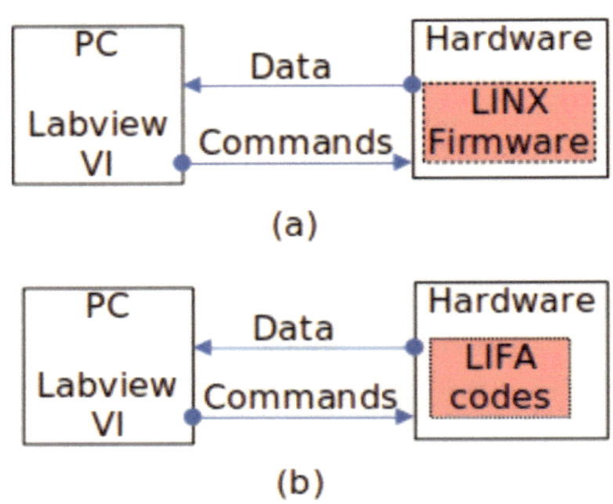

Figure 1.
The diagram of LINX system (a) and LIFA system (b).

memory, the diagram shows in **Figure 1(a)**. The LIFA users, instead of uploading the firmware as the LINX users do, edit the LIFA codes in IDE software (Integrated Development Environment) to function their system then upload to the hardware devices. For the graphical programming, both users have to download the packages and install them though VIPM (VI Package Manager), then the modules included in LINX and LIFA will appear in the menu windows of LabVIEW program. To communicate with the hardware devices, the LabVIEW graphical program can use the modules of read and write presented in the packages to upload commands to the hardware devices or download data to the PC.

Because of the independent projects to National Instrument, both LINX and LIFA occurred errors which caused by the compatibility with LabVIEW versions especially in LIFA applications, the common errors are 5002 and 5003 represented by the lost connection establishment. The LINX applications do not present the such errors and contains more various hardware firmware compatible with Arduino versions and Rashberry PI as well, however, the setup for sampling frequencies for data acquisitions are limited and much depended on the loop LabVIEW cycle. Therefore, depending on the requirements of applications, users can choose the model of LINX or LIFA for their embedded systems.

There are number of studies used LINX to apply Arduino module for low-cost embedded systems such as 3D object reconstruction with 3-axis scanning [9], temperature acquisition and control [10], DC motor control using PID (Proportional Integrated Derivative) controller [11], study transient response of RC circuit [12], an control the traffic signal [13]. Almost of the studies focused on the process control where the objects responses are limited in few milliseconds considering to few kilohertz sampling frequency. In the case of requirement for faster sampling frequency which has to be set on an optimum code to obtain, because of the fixed configuration, the LINX toolbox should not be chosen.

For the real-time applications using Arrduino and LabVIEW, a number of users considered LIFA as the toolbox for LabVIEW which toward embedded systems based on Arduino. Some typical applications such as a real-time simulation of the PV (Photo-Voltaic) panel [14], a ECG 12 leads (ElectroCardioGraphy) signal real-time observation [15], a digital stethoscope for body sound signal real-time collection [16], a design of real-time battery monitoring [17], Robotic arm real-time controller design [18], and a water level observer implementation [19]. Because of open source in LIFA and the users have to program the firmware procedures in Arduino code, the sampling frequency can be improved. Here is the reason that LIFA is preferred for professional users who are requested some skills in programming for hardware device such as micro-processors or micro-controllers.

For the LINX users, after the correspondent firmware chose and uploaded to the flash memory of the hardware devices, the graphical programs designed in LabVIEW response for all functions of the embedded systems. An example of the program is shown in **Figure 2** represents a typical analog signal acquisition using LINX. To initiate the program, a block named as "Open" addresses the in/out port of the connected hardware devices, here, is used of Arduino Uno which the port is defined as serial port through USB connection.

As the figure detailed, the block called "Analog Read" collects the signal from the channel corresponding to the analog input pin of the Arduino Uno module. The signal acquisition procedure is performed in the while loop until users stop it through the stop button or an error occurred. When the while loop is stopped, the block of "Close" terminate the program procedure. Here, to control the sample frequency a block named as "Wait" can be used. When the "Wait" block added with a constant or control parameter, the cycle of the while loop is controlled, hence, the sampling frequency is handled. Because of the dependency of the while loop cycle

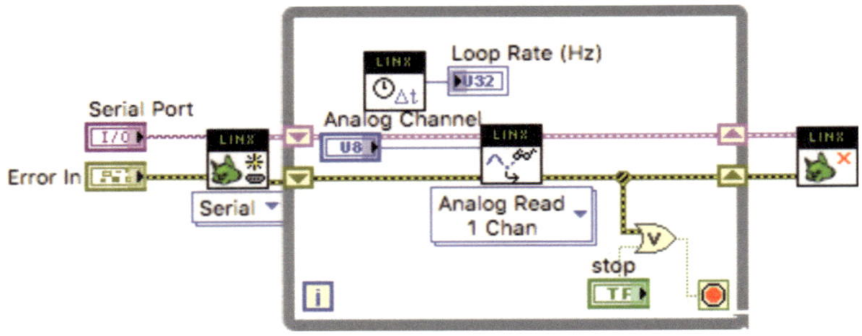

Figure 2.
Diagram of a typical acquisition LabVIEW program using LINX.

to the computer system cycle, the sampling frequency in the case of LINX using is not stable. Moreover, the hardware sampling frequency is set as constant in LINX firmware, therefore the embedded system sampling frequency should be slower than the value set in the firmware.

To use the LIFA, the applicants should compile a code through the IDE software. The mostly simple code can work with LabVIEW program called "LIFA_Base.ino" which usually packed in the LIFA package software. The codes are listed in **Figure 3** below:

```
#include <Wire.h>
#include <SPI.h>
#include <Servo.h>
#include "LabVIEWInterface.h"
 void setup()
{syncLV();}
void loop()
{checkForCommand();
if(acqMode==1)
{sampleContinously();}
}
```

Figure 3.
The codes written in LIFA_Base.Ino file.

where, the syncLV() is a function defined in LabVIEWInterface.h which provides a handshaking procedure with LabVIEW NI-VISA port configuration. For Arduino hardware devices, the USB port assigned as serial communication is used. When the connection with LabVIEW program is synchronized, both loops of the LabVIEW program and the Arduino firmware are operated. The loop in the Arduino firmware always check the code sent from LabVIEW then run the correspondent procedures otherwise the code of "sampleContinously()" is operated. The sampling procedure of the LIFA firmware requires a sampling frequency parameter to control the sampling cycle. Therefore, for the LabVIEW program, the users can completely control the sampling time which is one of the significant parameters for embedded systems especially in discrete control and data acquisition.

Figure 4 shows a typical acquisition LabVIEW program. The connection process initiated by the block called"Init". Throughout the initialization procedure, the parameters include baud rate, communication, Arduino board type, communication protocol, and data packet sized should be configured. When the connection is successful established, the analog signal is read via an analog input pin with

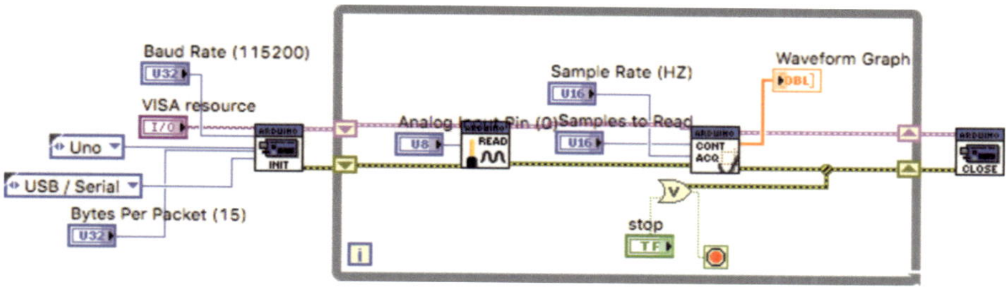

Figure 4.
Diagram of a typical LabVIEW program using LIFA.

the block of "Analog Read", then, the converted numerical data collected through the block named as "Continuous Acquisition" located on the Arduino menu in the popup programming menu. The data acquisition procedure requests parameters such as "samples to read" and "sample rate" to collect data with a certain sampling cycle and a pack of samples. The collected data is used to display with a graph panel called "Waveform Graph" originated in LabVIEW software. The data acquisition is stopped by the block named as "Close" to release the PC memory.

Unlike the procedure using LINX, the data acquisition program using LIFA does not depend on the loop cycle in PC since the sampling frequency is configured to the Arduino board and with requirement of each issues, the sampling time can be adjusted appropriately. However, the users have to code the program for hardware device to adapt with the LabVIEW program. Beside the hardware coding skills, the compatibility of LIFA firmware with LabVIEW is also a problem. With each version of LabVIEW software, some LIFA package version has bugs errors of connection such as 5002 and 5003 but this errors does not occurred while use of another LIFA package version. Therefore, the user should test their program with some LIFA package version to complete the embedded system.

3. Examples in signal processing

One of the application categories in use of embedded system is digital signal processing. The embedded system with numerical processes is much convenient than the analog or hardware process even if the low response time in the numerical system. LabVIEW is one of the options for embedded system development. The integration of LabVIEW program with hardware devices such as Arduino is great deal to reduce the cost. In this section, there are two examples are introduced: ECG signal processing and body sound signal processing.

ECG signal is a biomedical signal used to diagnose pathology related to cardiovascular disease. A simple ECG measurement system is designed with 3 leads placed on to human limbs to observe flutter wave and detect the complex waves include P, Q, R, S, T. To obtain the spatial signal of a heart, the electrodes should be placed on the chest with 6 leads assigned from v1 to v6. Here, with 4 electrode placed on all limbs, the 12 leads ECG measurement system is built. The 10 electrodes are used to monitor 12 signal of a cardiovascular system from the limbs to heart.

The low-cost hardware design for 12 leads ECG measurement based on Arduino Uno was reported [15]. To obtain the signal for storage and analysis, the signal is numerical converted and transferred to PC with software. For convenience, LabVIEW programming software was chosen. Because of the low spectrum range of ECG signal (0 to 100 Hz), the required sampling frequency is also low as well. The amplitude of ECG signal is weak (millivolts level), to observe the signal the amplitude have to be amplified with gain ranged from 30 dB to 40 dB, therefore, the ECG signal is contaminated with a number of the gained electromagnetic noises in the spectrum range.

To reduce the electromagnetic noises as well as increase SNR (signal-to-noise ratio) of the signal, beside the analog filters [15], digital filters is also considered since the difficulties of analog high-order filters design. After the signal purification, the wave of signal should be displayed then stored in PC. Therefore, the function diagram of a LabVIEW program contains signal acquisition, filtering, display, and storing. To compare the filtered signal and the raw signal, two waveform graph panel are used. The diagram has shown in **Figure 5**.

As **Figure 5** presented, the data is transferred to the data acquisition block and displayed in a graph plotting as well as filtered by the signal filtering. After the noise reduction processes, the filtered data is displayed and then stored in PC storage. Because of the low sampling frequency requirement, the LabVIEW-Arduino Uno interface can be chosen as LINX. By use of LINX package, the sampling cycle can adjusted through the "wait" block in LabVIEW to limited the while loop cycle. Typically, with the spectrum range of ECG signal, the maximum frequency reaches to 100 Hz, hence, the minimum sampling frequency should be 200 Hz as the Nyquist criteria requirement. However, the enlargement of sampling frequency provides better frequency resolution under frequency analysis.

The 12 leads ECG system contains 8 channels, includes V1 to V6, lead I and lead III. **Figure 6** shows the analog read program which was built with LINX and LabVIEW The data acquired through a block called "Analog Read N Channels", this block enable to read multiple channels simultaneously. Because the sampling cycle of the embedded system depends on the LabVIEW while loop cycle, therefore the

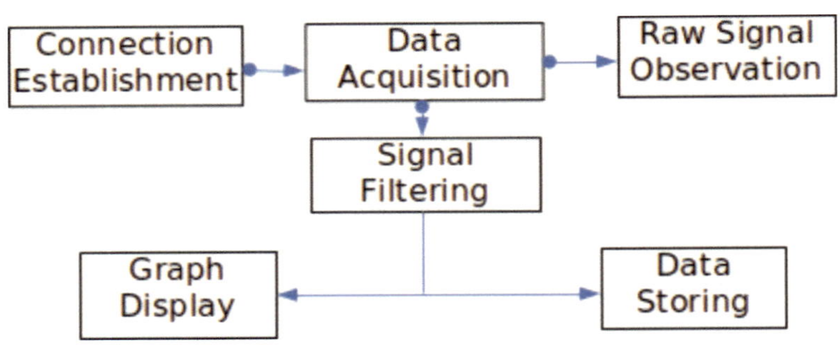

Figure 5.
Function diagram of the LabVIEW ECG signal acquisition.

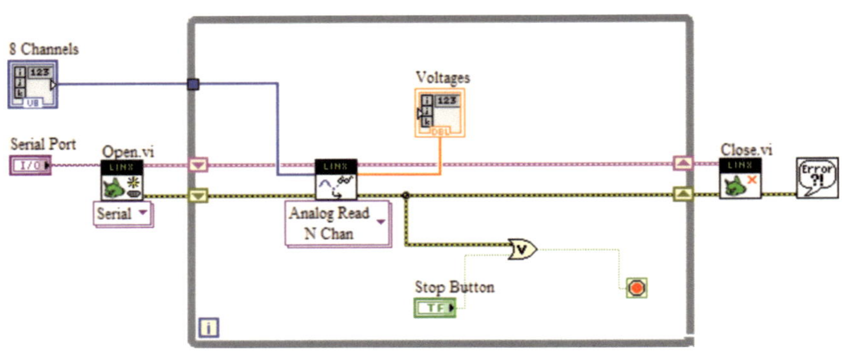

Figure 6.
The LabVIEW analog read program for 12 leads ECG measurement.

sampling time is set via "Time Delay" block. For this development, the sampling cycle is set with value of 0.003 seconds, corresponding to sampling frequency of 334 Hz approximately. The setting of the sampling cycle is shown in **Figure 7**.

Because of the multi channels reading, the data is packed in to a 1D array. To separate the data in each channel, a block named as "Index Array" was used. With each indexed number, a correspondent numerical value is extracted to an indicator. Based on the extracting in multi loops, the numerical values is arranged as a data 1D array which can be used to plot the raw data of each channel. The extraction program is indicated in **Figure 8**. Here, because the data indexed from 0, the channel number of 0 to 7 is assigned to extract the data of correspondent channels. The lead indicator named as the analog input pin corresponding to the electrode of the hardware device.

To reduce the noises from the collected data, a bandpass Butterworth filter was used. The noises contain both DC (Directed Current) components and AC (Alternative Current). In a signal waveform, AC components are the significant information that the observer can use for analysis. To eliminate the DC components in the ECG signal, a highpass filter with zero-close cut-off frequency should be addressed. To maximize the spectrum range, the zero-close cut-off frequency is chosen as 0.05 Hz while the high cut-off frequency of the low-pass filter is located at 100 Hz. The combination of the low-pass and the high-pass filter is the band-pass filter with the passed frequency range of 0.05 Hz to 100 Hz. The filter LabVIEW diagram is shown in **Figure 9**. Since the narrow range from stop frequency to cut-off frequency requests pitched edge, the filter order should be high. For this ECG spectrum range, the order of the band-pass filter is set as 6.

LabVIEW waveform graph block plots a waveform with 1D data typically in a linear variation of counted number of sample index. Therefore, the waveform

Figure 7.
The sampling cycle setting time with "time delay" block.

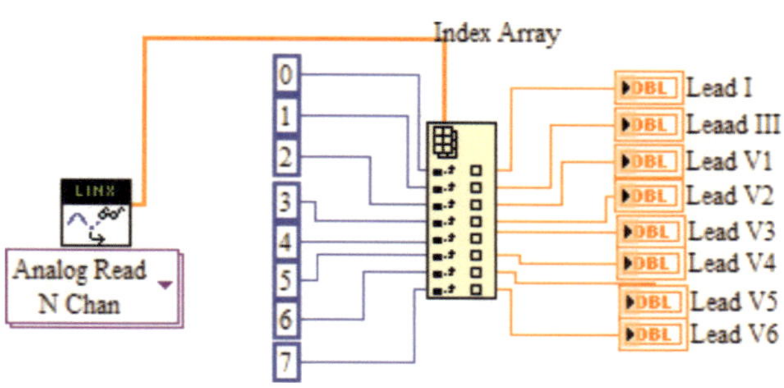

Figure 8.
The multi-channels data extraction program.

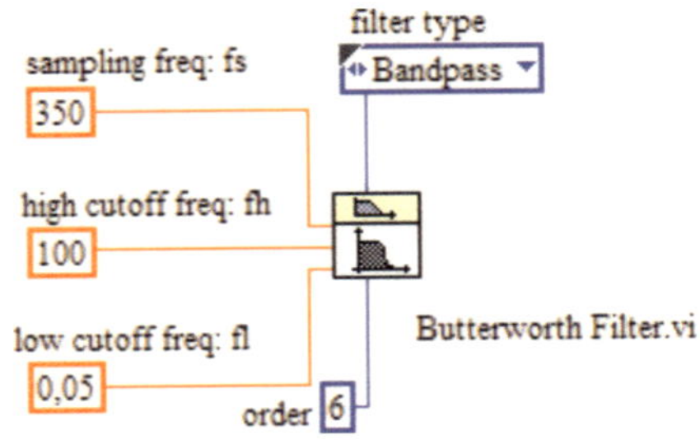

Figure 9.
LabVIEW diagram of the signal filtering program.

cannot be observed in the time-domain precisely. To indicate the graphs amplitude distribution in time-domain, the real-time clock can be combined with the ECG numerical data to be 2D data array, this process so called waveform building. Based-on the waveform building, the data plotted in a graph panel named as "Waveform Chart" in multiple channels display. The diagram of waveform building of amplitude data and real-time clock combination has shown in **Figure 10** while the array of the such 8 channel is presented in **Figure 11**.

The function of data storing can be designed with "creat file" block in "File I/O" menu. However, the data acquisition and file create in the same "while loop" may delayed the cycle of the loop. Therefore a structure Producer-Consumer which allow the parallel jobs; independent loops. By the structure, the received data in the first loop is enqueue then transmitted to the second loop named as dequeue and queue for data obtaining. The procedure is shown in **Figure 12**. With such program structure, the obtained data can be stored parallelly with the data acquisition

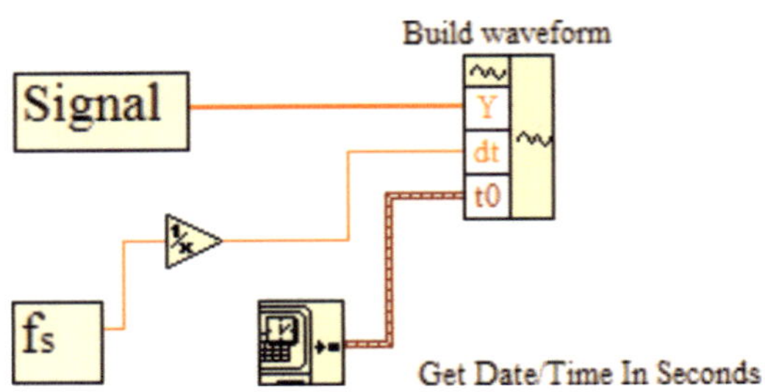

Figure 10.
The waveform building of a single channel amplitude data with the correspondent real-time clock data.

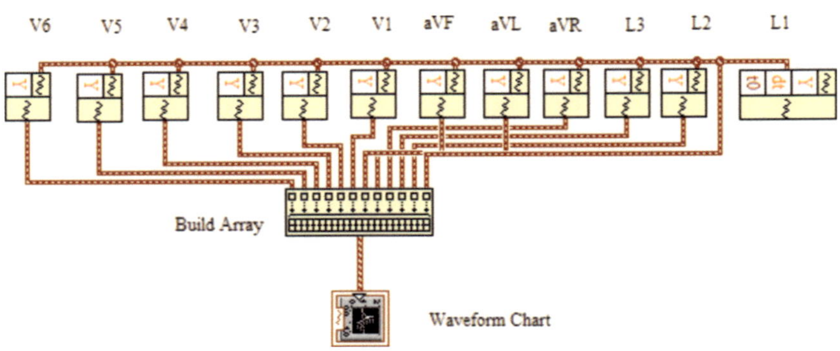

Figure 11.
The waveform building of 8 channels with amplitude and real-time clock data combination.

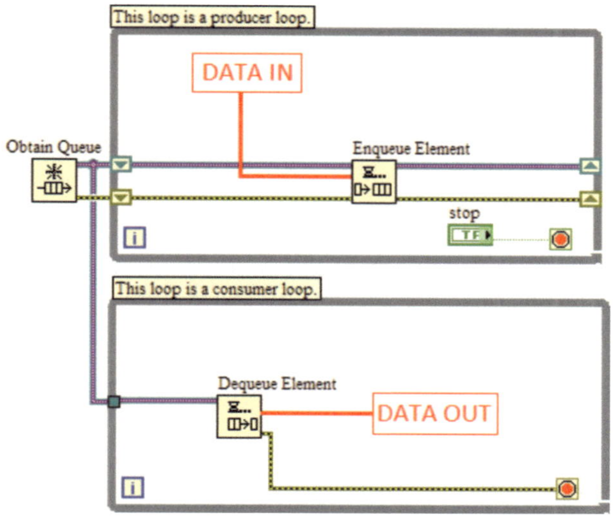

Figure 12.
The producer-consumer structure program for data storing.

procedure with the jobs of signal filtering, waveform building, graph plotting. This structure is also used in the case of that the variable need to be processed in multi loop arranged in the same program with synchronization requirement.

Another approach for data acquisition system with precise sampling frequency is LabVIEW applications with LIFA package. For LIFA users, as above discussion, the compatibility with each LabVIEW version should be checked since the NI-VISA of each LabVIEW version does not support for all LIFA packages, causing to error connection. Eventhough, the embedded system development community with LabVIEW still prefers LIFA because of simplicity and precision. With a requirement of fast sampling frequency, LINX might not be satisfied. In the second example, the observed objects are body sounds which their spectrum ranged in 0 to 20 kHz, the maximum frequency is much higher oscillation than the ECG signal. To observe the whole spectrum, at least, the minimum sampling frequency should be set at 44 kHz to satisfied Nyquist criteria. A test with this parameter based-on LINX application resulted the instability of the sampling frequency.

To obtain experiments of embedded systems collecting the body sound data such as heart sound and lung sound, a wide range stethoscope was designed [20]. The body sound is transduced to electromagnetic wave then converted to numerical data by Arduino board and read by a LabVIEW program. Instead of LINX, LIFA is used in this development. Similarly to the procedure has shown in **Figure 4**, the

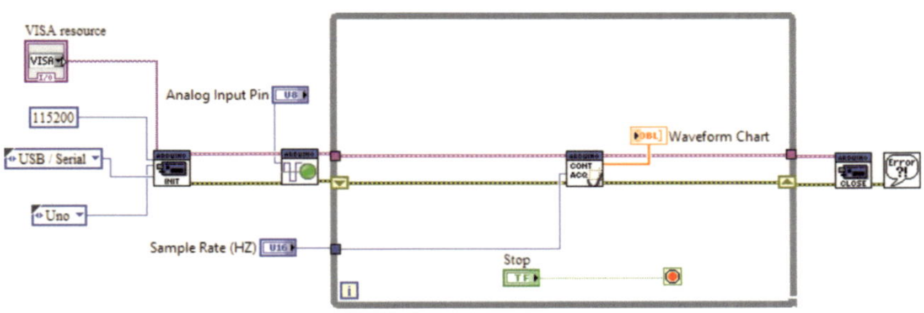

Figure 13.
The LabVIEW body sound data acquisition based on LIFA applications.

LabVIEW data acquisition program based-on LIFA is shown in **Figure 13**, where the block of "Analog read pin" is located outside of the while loop to avoid the replication of the setup pin mode for each cycles. Here, the sampling frequency can directly input to the "control" block as 88 kHz to satisfy the requirement of whole spectrum range observation. For the compatibility, this experiment was check with all current LIFA package with LabVIEW 2018 sp1. The result suggested that the package of 1.3.0.26 is great deal for this development without any errors.

Because of the conversion from the acoustic wave to the electromagnetic wave, the signal is contaminated with electromagnetic wave noises, hence, the noises should be reduced to enhance the SNR for the observation. To ensure the noise minimization, the digital band-pass filter is applied with cut-off frequencies are set at 5 Hz for low frequency cut and 35 kHz for high frequency cut, respectively. This block is different to the "Butterworth band-pass filter" which used in ECG acquisition system that the order of the filters are calculated based on the cut-off frequency and stop frequency parameters. The procedure application is much easier than the use of the filter block such as above Butterworth configuration. This block is packed in the toolkit named as "Digital Filter Design Toolkit". The filtering procedure diagram is shown in **Figure 14**.

During the sound data acquisition, beside the storage for future use, the sound as well as the quality can be checked directly by the hearing of users. To reconstruct the acoustic oscillation, the data conversed from the discrete format to analog wave through 24 bit DAC (Digital to Analog Conversion) sound card integrated in PC. The signal generates the body sound with a speaker with built-in amplifier. The functional procedure is shown in **Figure 15**, where the sound amplitude possibly adjusted with a digital amplifier called as "Set Volume" and generated to the output phone jack through the block of "write". Both blocks are located in Graphics and Sound menu.

Both applications based-on use of LINX and LIFA are introduced for examples to visualize the low-cost embedded system design with Arduino and LabVIEW.

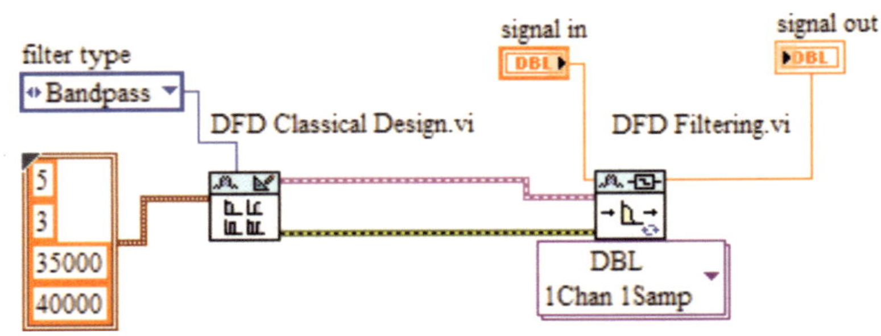

Figure 14.
The filter applying to the body sound signal in the LabVIEW program.

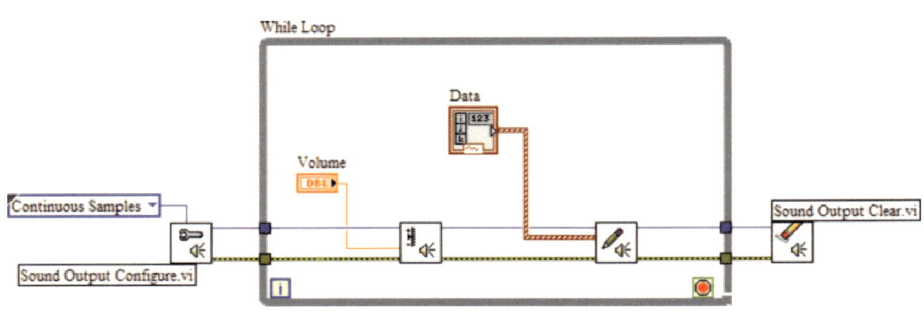

Figure 15.
The program with function of direct sound regenerating.

Even if the current drawback of the toolboxes are available, the development still can be solved in certain circumstances. With the experiment conditions and by the demonstration, these developments suggest another applications which toward the embedded system building without the traditional NI hardware devices. However, these development still need more work to complete the issues since the bug and error may occurred in some unknown conditions.

4. Another approaches

Beside Arduino boards, there are many commercial products are still developing such as Raspberry pi for remote access network or IoT (Internet of Things) system, chipKit for function enhancement, ESP8266 for wifi applications. Recently, to open a change for mobile AI (Artificial Intelligent) system, NVIDIA company released Jetson nanno kit which specified with quad-core CPU and 128-core GPU. Almost of current embedded hardware devices use USB connection with asynchronous communication protocol for data exchange. This means with the definition of NI-VISA which support for the hardware using USB port, the commands and the data can be full-duplex transferred between PC and the embedded hardware devices. The requirement for LabVIEW interface can be solved through firmware library programming.

In the case of large amount of logic components integration required such as parallel filters, numerical array convolution, the embedded system FPGA-based (Field Programmable Gate Array) with LabVIEW can be considered. Unlike the LabVIEW interface package such as LINX and LIFA, the FPGA-based embedded system has to be originated by NI hardware devices or use of other hardware compatible with NI-myRIO platform. With the optimization of logic circuit with specified functions or procedure, the fast response of the FPGA-based embedded systems possibly satisfy the real-time requirement. An example of such application was reported in 2010 with real-time response for DC motor controller design [21].

The advantage of LabVIEW embedded systems that is the enriched update of libraries and block module in each version which support for data processing such as signal processing, image processing, network routing, control system, computer vision, IoT, and more. In large scale system level, when the real-time characteristics is required, the distributed system with multiple computers network possibly the solution. Therefore, for the LabVIEW embedded system in the near future, a trend of functionalizing hardware device is potential development. With the standardization of the data transfer protocol, the computer peripheral design based on embedded system will focus on the firmware and software design rather than the hardware design as before.

With the development of the operating system for smart devices such as Android and iOS, NI LabVIEW also provide a solution that the smart devices users can observe acquired data and control system objects distantly through a service called WebVI, which means the embedded systems in the future can be observed and control through smart devices such as smartphone, tablet when they are connected to internet, an IoT solution of LabVIEW application development [22]. Based-on this enhancement, the category of applications toward IoT system can be widely developed in the near future.

5. Conclusions

In this chapter, embedded systems based-on LabVIEW applications with low-cost hardware devices are introduced and discussed. For the application developments, the topic of LabVIEW interface with hardware device is focused with use of LIFA and LINX package. The similarity and the difference between both interface package is also discussed and analyzed. To visualize the embedded system design process LabVIEW-based, two experiments in data acquisition and signal processing with use of LINX and LIFA package were introduced as well. LIFA package provide the simplicity in use but remain the problem of compatibility with each LabVIEW version while LINX package is wide range of hardware devices support but some significant characteristics are very difficult to configure. To adapt with a number of modern smart devices currently, LabVIEW also open a service to support IoT application category. This support promises the potential development in the near future.

Acknowledgements

This work is contributed by students who used to study in Biomedical Engineering Laboratory belonging to Hanoi University of Science and Technology. The author also appreciate the editor and secretary for encouragement to complete this chapter during the Covid-19 epidemic.

Author details

Trinh Quang Duc
Hanoi University of Science and Technology, Hanoi, Vietnam

*Address all correspondence to: duc.trinhquang@hust.edu.vn

References

[1] LabVIEW. Available from: https://en.wikipedia.org/wiki/LabVIEW [Accessed: 2021-04-15].

[2] Arduino. Available from: https://en.wikipedia.org/wiki/Arduino [Accessed: 2021-04-15].

[3] NodeMCU. Available from: https://en.wikipedia.org/wiki/NodeMCU [Accessed: 2021-04-15].

[4] David Kushner. The Making of Arduino. IEEE Spectrum. 26 Oct. 2011.

[5] LIFA (LabVIEW Interface For Arduino). Available from: https://github.com/labviewhacker/lifa [Accessed: 2021-04-15].

[6] Marco Schwartz and Oliver Manickum. Programming Arduino with LabVIEW. Packt Publishing; 27 Jan, 2015.

[7] LINX: How LINX Works. Available from: https://blog.digilentinc.com/how-linx-works/ [Accessed: 2021-04-15].

[8] NI-VISA: Overview. Available from: https://www.ni.com/en-vn/support/documentation/supplemental/06/ni-visa-overview.html [Accessed: 2021-04-15.

[9] Rob A, Panoiu C. Using LabVIEW Linx for Creating 3D Objects. IOP Conf. Series: Materials Science and Engineering; 447(2019) 012027. DOI: 10.1088/1757-899X/477/1/012027.

[10] Yusuf Ayuba. Temperature Control and Data Acquisition Method for Factory using LabVIEW. International Journal of Computer Engineering and Technology; 2016, 7(2), pp. 01-14.

[11] G. Gaşparesc. PID control of a DC motor using LabVIEW Interface for Embedded Platforms. *12th IEEE International Symposium on Electronics and Telecommunications (ISETC)*, Timisoara, Romania, 2016, pp. 145-148, DOI: 10.1109/ISETC.2016.7781078.

[12] N Suwondo and D Sulisworo. Hands-on Learning Activity Using an Apparatus for Transient Phenomena in RC Circuit Based-on Arduino UNO R3-LINX-LabVIEW. International Journal of Online and Biomedical Engineering. 2017, 13(1), pp. 116-124.

[13] Mihai Bogdan. Traffic Light Using Arrduino Uno and LabVIEW. The 12th International Conference on Virtual Learning ICVL 2017; 2017, pp. 287-290.

[14] A El Hammoumi, S Motahhir, A Chalh, A El Ghzizal and Derouich. Realtime virtual instrumentation of Arduino and LabVIEW based PV panel characterristics. International Conference on Renewable Energies and Energy Efficiency (REEE2017). IOP Conf. Series: Earth and Environmental Science 161 (2018) 012019 DOI:10.1088/1755-1315/161/1/012019.

[15] Son Nguyen Van, Duc Trinh Nguyen, Giang Nguyen Hoai. Development of a Low-Cost Arduino-Based 12-Lead ECG Acquisition System And Accompanied LabVIEW Application. International Journal of Engineering and Advanced Technology. 2019, 09, pp. 1641-1648.

[16] Son Nguyen Van, Duc Trinh Quang, Giang Nguyen Hoai, Quynh Nguyen Thi Huong, Khanh Pham Xuan. Software Design Collection and Handling of Signal Sound Body, International Journal of Engineering and Advanced Technology. 2019, 09, pp. 260-262.

[17] A. Jamaluddin, L. Sihombing, A. Supriyanto, A. Purwanto and M. Nizam. Design real time Battery Monitoring System using LabVIEW Interface for Arduino (LIFA). *2013 Joint International*

Conference on Rural Information & Communication Technology and Electric-Vehicle Technology (rICT & ICeV-T), Bandung, Indonesia, 2013, pp. 1-4, doi: 10.1109/rICT-IceVT.2013.6741525.

[18] B. Harish, U. Jyothsna. LabVIEW Interface with Arduino Robotic ARM. International Journal of Science and Research. 2015, 4(11), pp. 2423-2426.

[19] K Premsagar, M. Akhila, P. Prashanthi Reddy, P Srinivas. Unused Water Level Observing Structure Implementation Using LabVIEW and Arrduino. Journal of Critical Reviews. 2020, 7(12), pp. 1344-1350.

[20] Trinh Quang Duc, Nguyen Van Son, Nguyen Hoai Giang. A Developed Design to Enlarge the Spectrum for Analog Electronic Stethoscope. Journal of Engineering Science and Technology Review. 2019, 12, pp. 179-184.

[21] F. H. Ali, M. M. Hussein and S. M. B. Ismael. LabVIEW FPGA implementation of a PID controller for D.C. motor speed control. 2010 1st International Conference on Energy, Power and Control (EPC-IQ). 2010, pp. 139-144.

[22] LabVIEW support for iOS and Android Operating Systems. Available from: https://knowledge.ni.com/KnowledgeArticleDetails?id=kA00Z0000019NNjSAM&l=en-VN [Accessed: 2021-04-15].

Chapter 6

Certain Applications of LabVIEW in the Field of Electronics and Communication

Prema Ramasamy, Shri Tharanyaa Jothimani Palanivelu and Abin Sathesan

Abstract

The LabVIEW platform with graphical programming environment, will help to integrate the human machine interface controller with the software like MATLAB, Python etc. This platform plays the vital role in many pioneering areas like speech signal processing, bio medical signals like Electrocardiogram (ECG) and Electroencephalogram (EEG) processing, fault analysis in analog electronic circuits, Cognitive Radio(CR), Software Defined Radio (SDR), flexible and wearable electronics. Nowadays most engineering colleges redesign their laboratory curricula for the students to enhance the potential inclusion of remote based laboratory to facilitate and encourage the students to access the laboratory anywhere and anytime. This would help every young learner to bolster their innovation, if the laboratory environment is within the reach of their hand. LabVIEW is widely recognized for its flexibility and adaptability. Due to the versatile nature of LabVIEW in the Input- Output systems, it has find its broad applications in integrated systems. It can provide a smart assistance to deaf and dumb people for interpreting the sign language by gesture recognition using flex sensors, monitor the health condition of elderly people by predicting the abnormalities in the heart beat through remote access, and identify the stage of breast cancer from the Computed tomography (CT) and Magnetic resonance imaging (MRI) scans using image processing techniques. In this chapter, the previous work of authors who have extensively incorporated LabVIEW in the field of electronics and communication are discussed in detail.

Keywords: speech signal, embedded systems, robotics

1. Introduction

The recognition and processing of speech signal plays a key role in any real time implementation. The plethora of approaches has been followed to analyze the continuous speech spectrum namely Perceptual Linear Prediction (PLP) statistical approach, Relative Spectral (RASTA) method, Mel-frequency cepstral analysis (MFCC), hidden Markov models (HMM), artificial neural networks (ANN), etc. The speech signal processing can be further enhanced by filtering out the noisy environments using single channel or multi-channel methods.

Radek et al., [1, 2] have incorporated the multi-channel methods using least mean square algorithm (LMS) and independent component analysis (ICA) to provide the mathematical calculations to avoid additive noise in speech signal. For smart home implementation, the LabVIEW SW tool is used for visualization, speech recognition, virtual cable connection to the sound card, and the actual mathematical calculations within additive noise canceling. The speech recognition engine will provide communication between LabVIEW and the recognizer. Its function is to convert input command into stream of text outputs. The output will be compared with the recognized command and the appropriate command will be given to ON/OFF the devices like Fan, Television, Washing Machine, Vacuum Cleaner, Dish Washer, etc. in the home. The Virtual Audio Cable (VA – Cable) is a software used for communicating between the devices using the audio streams.

LabVIEW is capable of executing/implementing many parallel loops on FPGA and real-time controllers by using a set of complex math functions. To make it more powerful, it has been integrated with LabVIEW SoftMotion Module and Kollmorgen AKD drives and motors to control the smart machines in all aspects. It also helps in the maintenance of smart machines after their deployment in the world by implementing centralized resource management, diagnostics, and remote updates [3].

2. Speech signal processing

The speech recognition and processing are playing a vital role in various industries including data acquisition, hands-free mobile manufacture for physically challenged, instrumentation field and machine control. For implementing the speech recognition system in LabVIEW platform, the prerequisites include sound card, microphone and other accessories related to the hardware PC/Mobile. In the design of intelligent voice system using LabVIEW [4], Virtual Instrument (VI) can be incorporated to simplify the process of Voice Recognition Technology (VRT). It has been enriched with the features of graphical-based programming, digital data stream and simplified programming model to involve the audio front panel. Even though speech signal of different people may have similar power spectrum, the low frequency band will differ uniquely for each and every user. Therefore, after recording the voice it can be analyzed with the help of sound acquisition module in LabVIEW [4]. Once speech signal is received from the microphone, it is preprocessed and filtered before adding the Hamming window to calculate the power spectrum of each user. From the power spectrum, low frequency peaks can be obtained in order to identify the voice characteristics of individual user. Using Labview PSD module peak frequency is extracted from the spectrum of voice signal.

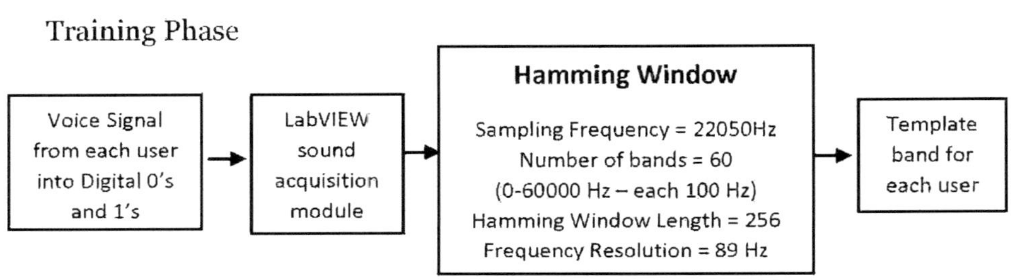

Recognition Phase

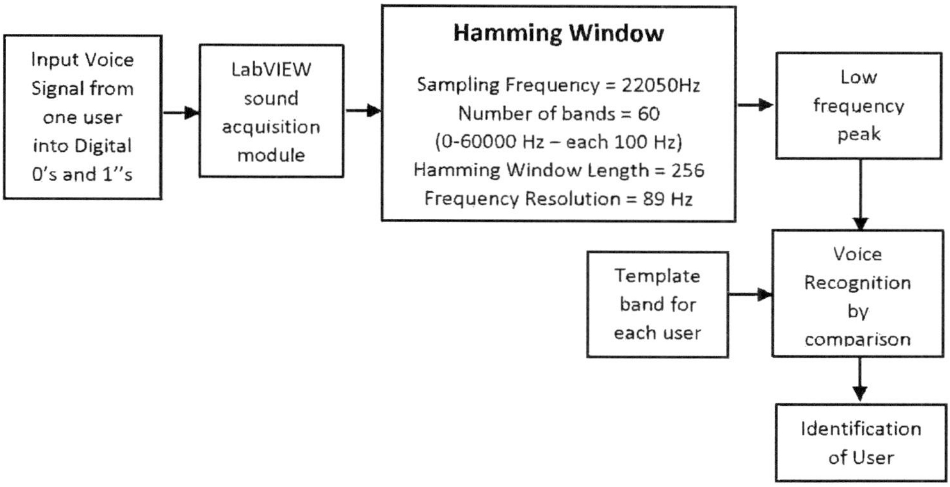

The design can be divided into training and recognition. In the training phase each user's voice is acquired from the input speech device, such as microphone. LabVIEW stores the sound as a waveform arrays. Each waveform in the array represents one channel. Therefore, for stereophonic system two waveform elements will be present in an array. Since each channel plays simultaneously in the stereo system, and the data will be represented in the form of Pulse Code Modulation (PCM). In the pulse code modulated data, each element in an array will be proportional to the pitch or intensity of the signal. Each array will be arranged from minimum to maximum with zero as centered. If the array data type is of 16-bit or 32-bit signed integer, then the values will range from 0 to 65536 for 16-bit which will be centered at 32768 similarly from 0 to 4,294,967,296 for 32-bit will be centered at 2,147,483,648. A buffer of elements each with the value of zero (32768–16 bit, 2,147,483,648–32 bit) represents silence. The recommended hardware for sound and vibration measurement in the category of high performance model includes NI 4461 and NI 4462, high density model includes NI 4495, NI 4496, and NI 4498, low cost model includes NI 4472 and NI 4474 and portable/compact model includes NI 9233, NI 9234, NI 4432 and NI 4431. Using the sound acquisition module the voice signal will be read through PC sound card. The power spectrum will be obtained through hamming window and after identifying the low frequency peaks template for each and every user will be stored in the module. In the recognition phase, the voice of any user can be taken and it will be compared with the template frequency band to verify the identity of concerned person.

3. Labview for embedded systems

LabVIEW, an easier graphical programming environment than GUI based application development platforms like Visual Basic and Visual C++. The conventional instrumentation does not perform analysis and controlling of multiple signals at the same time but LabVIEW with Virtual Instrumentation will do so. The main advantage of LabVIEW is that it can be used for PC-based monitoring and data logging.

The latest processors that have been used for the embedded system design are ARM microcontrollers. LabVIEW has a comprehensive graphical development environment module for ARM microcontrollers. This has been jointly developed by Keil - An ARM Company and NI instruments for Embedded System design applications. This module will reduce the development cost and fasten the programming part for Embedded System Design. This module includes hundreds of analysis and signal processing functions, integrated I/O, and an interactive debugging interface [5]. These advantages of LabVIEW pays a path for its usage in embedded applications for controlling. There are many case studies for such kinds of applications.

The photovoltaic panel emulator can be designed using LabVIEW interfaces to identify the characteristics of electrical parameters of PV panel. In the model designed by Gurkan et al. [6], four PV panels has been monitored with the help of PV panel emulator and electronic potentiometer. All the panels and modules has been connected to the ATMEGA2560 main control unit. Because of the flexibility and enriched features, LabVIEW is incorporated to provide communication between PC and the main control unit. The serial communication port COMPORT in LabVIEW will be used to send array of strings to and from PC and PV panels. Using the LabVIEW specific interface can be designed to observe the electrical characteristics of PV panel in real time by adjusting the resistance values of electrical potentiometer.

A smart turf harvesting machine (ProSlap 155) has been designed using a CompactRIO controller by FireFly in 2014. The CompactRIO has been used as hardware for the machine and LabVIEW used as software. To perform the parallel operations, the FireFly has combined traditional fluid power systems with a servo-electric system on the machine using the LabVIEW and CompactRIO. The LabVIEW has been used to program 80 different sensors and 100 digital outputs of the hardware and provide a secure Internet gateway to remotely monitor and control the turf harvester. The ProSlap 155 turf harvester increases the harvesting speed to 20% and reduces the fuel consumption up to 50% and this increases farm productivity [7].

LabVIEW is one of the best software used in the production process control system applications. It has been used to remotely control the Crude Oil Separator in Crude oil industries. Wireless connectivity through TCP/IP protocol has been established between the LabVIEW and the PIC microcontroller for controlling and automation of crude oil separators. It has used to control the pressure and level of a Gas and Liquids in the crude oil separator (**Figure 1**) [8].

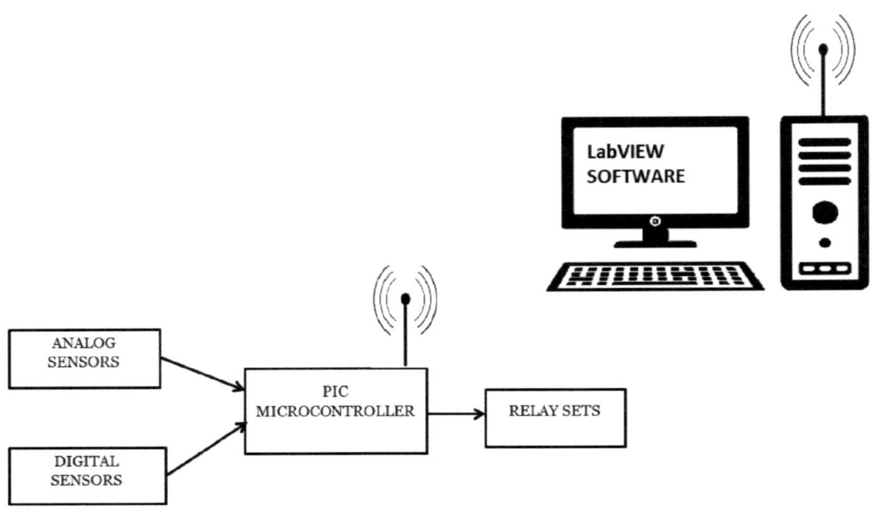

Figure 1.
System Block Diagram of Crude Oil Separator.

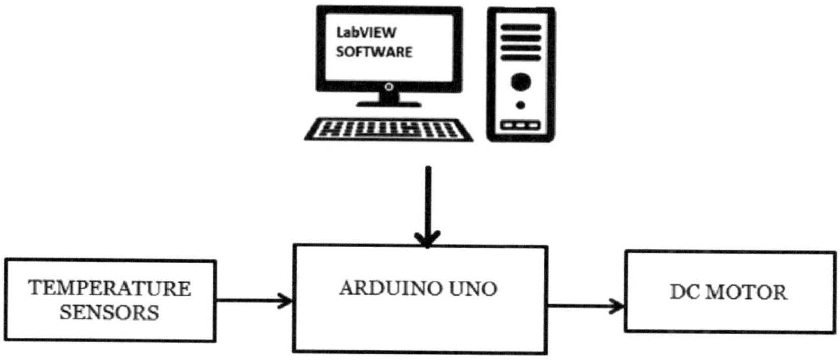

Figure 2.
System Block Diagram.

LabVIEW is also used to control a device. In this case study, the LabVIEW is used to control the speed of the DC motor based on the temperature in the devices like CPU, Air Conditioner, etc. The temperature sensor is used to measure the temperature in the environment of the system. The DC motor is connected with the Arduino board. The board will send the PWM output to control the speed of the motor. The duty cycle of the PWM output is decided by the LabVIEW based on the temperature value from the sensor. As the temperature varies, the speed of the motor will also be varied (**Figure 2**) [9].

4. Applications of LabVIEW in robotics and VLSI

Present world is looking for high throughput fast processing system with features that meets ideal specifications. As we know, a system complexity increases consequently the system hardware and software must be complicated to attain the desired target within the time specification. As we are going through the era of artificial intelligence and machine learning so that it is necessary to add technological improvements in our system designs to compete with latest market trends. In the field of robotics, software and hardware have equal responsibility and functionality to attain the accuracy while take part in a certain action. In order to create a stable perfect system, the designer can choose graphical based design rather than conventional system design.

LabVIEW can easily use as a flexible environment for the robotic design. LabVIEW is a development environment that will provide flexibility for automated product design validation and monitoring of a machine status. Here the designer can use the language "G" for the system coding and drag and drop option of appropriate functional blocks will generate a working system according to the specifications.

Industrial machine builders [10] are forced to deliver machines that meets several demands. It is necessary to hold their unique position in the respective field with respect to the competitors. LabVIEW is mainly used for building a graphical user interfaces for systems and also it is utilizing inherent parallelism for instrumentation, control system and robotics. If the designers are looking for a simple design in the hardware and software LabVIEW is the best option for that. Robotic arm control mechanism needs more accuracy to work in the fields such as of medical electronics and chemical industry. The processor will run according to the program coded using lab view and the processor will generate the control signals to control and monitor the movement of robotic arm.

LabVIEW is applicable for standalone as well as moving system design. Wherever the application of a mixed system comes then the design complexity will increase and monitoring will be more complicated. Here the graphical user interface [11] will be more useful to control the action of the system and its programming.

5. Medical applications

Jung et al. [12], integrated the flexible multimodal sensor based E-Skin which will mimic the skin in human beings. The sensors like temperature, pressure, flow and humidity everything has been included to create the characteristics of human skin. To measure the varying resistance in the hair sensor they customized the PC-Controlled Servomotor LabVIEW Tower Pro SG 90 and PC-recordable acquisition board NI USB-6218. The E-Skin plays vital role in soft electronics to enhance the Human-Machine Interface (HMI) for analyzing the ECG, EMG and other parameters. The wearable and flexible electronics [13] paves the way for comfort in using these devices.

6. Education

Many real time remote –based laboratory models have been constructed to help the students in schools and higher education institutions to perform the research and gain the knowledge from their own location itself. Several laboratories including the light sensing experiment designed by Singh et al. [14] using LabVIEW, LDR and LM35 sensor, the phasor estimation algorithm using NI ELVIS kit designed by Mondal et al. [15], operational process monitoring application using SHRIMP VL application, analog electronics experiment using “RedPitaya” [16] etc. The LabVIEW has been effectively and efficiently included in many studies in literature to enhance and boost the research community. Polat et al. [17], designed the Distance Vocation Education and Traning D-VET to monitor the lighting levels of LED lamps in remote laboratory with the help of PID controller, Sensor, Actuator, NI-IMAQ software and CompactRIO DAQ.

Azenova et al. [18] designed an e-learning course for second year industrial students under the category of laboratory experiments in electronics. The LabVIEW is incorporated in the Goodyear e-learning model they have adopted for implementing to their students. The basic laboratory experiments like clipper, clamper, inverting and non-inverting amplifier, summing and differential amplifier were simulated with the help of virtual instruments front panel in LabVIEW.

7. Miscellaneous applications

The fault detection in analog circuit can be easily analyzed with the help of LabVIEW interfaces using deep forest learning method. Zhen et al. [19] have designed the fault diagnosis scheme with the Waveform generator Agilent 33250, Data acquisition module NI 1042q, LabVIEW, Digital oscilloscope Agilent 54853 along with power supply and testing circuit. The outputs of testing circuit are collected by data acquisition module and stored in the LabVIEW and used for further diagnosis.

For assessing the quality of water to ensure the living of aquatic species like fishes in a healthy environment. Othman et al. [20], proposed a system using the sensors to measure the pH level and temperature of the water in the tank. The data collected from the sensors are given to the Data Acquisition System (DAQ) NI myRIO-1900. Then the data will be analyzed periodically in the workstation to ensure the safety of aquatic animals. The interface between DAQ and workstation is created with the help of LabVIEW software. The temperature and pH measurements are displayed in the front panel of LabVIEW.

Ahmed et al. [21] discussed the implementation of LabVIEW software to control the FESTO MPS PA compact workstations as part of remote laboratory. In the paper they designed the Easyport USB data acquisition card based on activeX elements to interface with LabVIEW and control the applications in Fluid Lab. To obtain the real time information from the Sensors and actuators of MPS-PA station various VI were included. To incorporate remote monitoring web server of LabVIEW is required. With the help of the VI the user can adjust the parameters of Controller and with help of IP camera and web server the MPS PA compact workstation can be monitored remotely. Proportional-Integral-Derivative (PID) and ON/OFF controllers these two experiments as part of laboratory course were implemented using the system designed by Ahmed et al.

Mikhoylov et al. [22] aimed to monitor the specific content from the broadcast channels using the help of Cognitive Radio (CR). For the real time data analytics, particular subset of channels from the selective frequency range will be chosen and the significant information in the subset will be relayed for monitoring in the remotely using the VI from LabVIEW. For monitoring Software Defined Radio (SDR) technology USRP 2900/2901 and LabVIEW were utilized in the design. The key factor in the application is to either receive & process or transmit the subset of signals in the digitized format with the help of the packages available in the LabVIEW.

8. Conclusion

The applications of LabVIEW in electronics and communication field has been presented in this chapter. The role of LabVIEW in Speech signal processing, Photovoltaic cell emulator, Fault diagnosis scheme, Cognitive radio signal analysis, FESTO workstation control, Water tank surveillance, Embedded system like the Crude Oil Separator, smart turf harvesting machine and robotics applications were discussed briefly. The demonstration and exploration of laboratory courses for school and college students are in reach of hand because of the interface provided by the LabVIEW and source of the Internet.

Author details

Prema Ramasamy*, Shri Tharanyaa Jothimani Palanivelu and Abin Sathesan
Bannari Amman Institute of Technology, Erode, Tamil Nadu, India

*Address all correspondence to: prema@bitsathy.ac.in

References

[1] R. Martinek, R. Kahankova, P. Bilik, J. Nedoma, M. Fajkus, and M. Skacel, "Speech signal processing using microphones NI 9234 and LabVIEW," ACM Int. Conf. Proceeding Ser., pp. 9-13, 2018, doi: 10.1145/3177457.3177501.

[2] R. Martinek, J. Vanus, J. Nedoma, M. Fridrich, J. Frnda, and A. Kawala-Sterniuk, "Voice communication in noisy environments in a smart house using hybrid lms+ica algorithm," Sensors (Switzerland), vol. 20, no. 21, pp. 1-24, 2020, doi: 10.3390/s20216022.

[3] "https://www.ni.com/en-in/shop/labview/how-do-i-use-labview-to-design-smart-machines.html." .

[4] Q. K. Xu, Z. Q. Du, and J. He, "The Design of the Intelligent Voice Recognition System Based on Labview," Appl. Mech. Mater., vol. 734, pp. 383-386, 2015, doi: 10.4028/www.scientific.net/amm.734.383.

[5] "Manual:Getting Started with the LabVIEW Embedded Module for ARM."

[6] S. Gürkan, M. Karapınar, H. Sorgunlu, O. Öztürk, and S. Doğan, "Development of a photovoltaic panel emulator and LabVIEW-based application platform," Comput. Appl. Eng. Educ., vol. 28, no. 5, pp. 1291-1310, 2020, doi: 10.1002/cae.22302.

[7] "https://www.ni.com/en-in/innovations/case-studies/19/smart-turf-harvesting-machine-boosts-productivity-and-reduces-cost.html." .

[8] A. S. Allahloh, S. Y. Sondkar, and S. Mohammad, "Implementation of online fuzzy controller for crude oil separator industry based on internet of things using LabVIEW and PIC microcontroller," *2018 Int. Conf. Comput. Power Commun. Technol. GUCON 2018*, pp. 341-346, 2019, doi: 10.1109/GUCON.2018.8675035.

[9] K. R. Asha, P. Suhada Tasleem, A. V. Ravi Kumar, S. Mallikarjuna Swamy, and K. R. Rekha, "Real Time Speed Control of a DC Motor by Temperature Variation Using LabVIEW and Arduino," *Proc. - 2017 Int. Conf. Recent Adv. Electron. Commun. Technol. ICRAECT 2017*, pp. 72-75, 2017, doi: 10.1109/ICRAECT.2017.50.

[10] C. Xiaowei and S. Chuanwei, "The study of robot simulation applications based on the LabVIEW robotics," in *2017 4th International Conference on Information, Cybernetics and Computational Social Systems (ICCSS)*, Jul. 2017, pp. 66-71, doi: 10.1109/ICCSS.2017.8091386.

[11] W. Banas, G. Cwikła, K. Foit, A. Gwiazda, Z. Monica, and A. Sekala, "Modelling of industrial robot in LabView Robotics," IOP Conf. Ser. Mater. Sci. Eng., vol. 227, no. 1, 2017, doi: 10.1088/1757-899X/227/1/012011.

[12] M. Jung *et al.*, "Flexible multimodal sensor inspired by human skin based on hair-type flow, temperature, and pressure," *Flex. Print. Electron.*, vol. 5, no. 2, 2020, doi: 10.1088/2058-8585/ab8073.

[13] Z. Rao, F. Ershad, A. Almasri, L. Gonzalez, X. Wu, and C. Yu, "Soft Electronics for the Skin: From Health Monitors to Human–Machine Interfaces," Adv. Mater. Technol., vol. 5, no. 9, pp. 1-27, 2020, doi: 10.1002/admt.202000233.

[14] A. K. Singh, S. Chatterji, S. L. Shimi, and A. Gaur, "Remote Lab in Instrumentation and Control Engineering Using LabVIEW," *Int.* J. Electron. Electr. Eng., vol. 3, no. 4, 2014, doi: 10.12720/ijeee.3.4.297-304.

[15] S. Mondal, C. Murthy, D. S. Roy, and D. K. Mohanta, "Simulation of Phasor Measurement Unit (PMU) using

labview," in *2014 14th International Conference on Environment and Electrical Engineering*, May 2014, pp. 164-168, doi: 10.1109/EEEIC.2014.6835857.

[16] C. J. Garcia-Orellana, M. Macias-Macias, H. Gonzalez-Velasco, A. Garcia-Manso, and R. Gallardo-Caballero, "Remote laboratory experiments of Analog Electronics based on 'RedPitaya,'" in *2016 Technologies Applied to Electronics Teaching (TAEE)*, Jun. 2016, pp. 1-7, doi: 10.1109/TAEE.2016.7528244.

[17] Z. Polat and N. Ekren, "Remote laboratory trends for Distance Vocational Education and Training (D-VET): A real-time lighting application," Int. J. Electr. Eng. Educ., pp. 1-16, 2020, doi: 10.1177/0020720920926679.

[18] A. Asenova, K. O. Street, and K. O. Street, "Design of university e-course in Electronics for future engineers English Language Faculty of Engineering Faculty of Telecommunications," vol. 5, pp. 5-9, 2020.

[19] Z. Jia, Z. Liu, Y. Gan, C. M. Vong, and M. Pecht, "A Deep Forest Based Fault Diagnosis Scheme for Electronics-Rich Analog Circuit Systems," IEEE Trans. Ind. Electron., vol. 0046, no. c, pp. 1-1, 2020, doi: 10.1109/tie.2020.3020252.

[20] N. A. Othman, N. S. Damanhuri, M. A. Syafiq Mazalan, S. A. Shamsuddin, M. H. Abbas, and B. C. Chiew Meng, "Automated water quality monitoring system development via LabVIEW for aquaculture industry (Tilapia) in Malaysia," *Indones. J. Electr. Eng. Comput. Sci.*, vol. 20, no. 2, pp. 805-812, 2020, doi: 10.11591/ijeecs.v20.i2.pp805-812.

[21] S. A. Ahmad, S. K. Alhayyas, M. A. Almansoori, N. A. Almenhali, F. S. Alsudain, and A. H. Alkhaldi, "Remote control of the FESTO MPS PA compact workstation for the development of a remotely accessible process control laboratory," Int. J. online Biomed. Eng., vol. 16, no. 5, pp. 84-103, 2020, doi: 10.3991/IJOE.V16I05.12809.

[22] V. Y. Mikhaylov and R. B. Mazepa, "Research of cognitive radio technology application cases in the tasks of providing information needs," *2020 Wave Electron. its Appl. Inf. Telecommun. Syst. WECONF 2020*, 2020, doi: 10.1109/WECONF48837.2020.9131163.

Section 2

LabVIEW in Modeling

Chapter 7

Advanced Modeling of Single Degree of Freedom System for Earthquake Ground Motion Using LabVIEW Software

R.B. Malathy, Govardhan Bhat and U.K. Dewangan

Abstract

In this paper, the structural responses at discrete time steps are evaluated to understand the linear dynamics characteristics of a structural system using LabVIEW (Laboratory Virtual Instrument Engineering Workbench) tool. Time History Analysis (THA) which is an essential procedure to design a reliable structure when the structure is subjected to dynamic loading is taken into consideration for the study. Direct integration method was used to find out the dynamic response of the structure as it is applicable for both linear as well as nonlinear range. Block diagram that perform step-by-step integration to analyze the linear single degree of freedom (SDOF) system has been prepared in LabVIEW. The processing of data is carried out till the equilibrium is satisfied at all discrete time points within the interval of solution instead of any time t. Different ground motion time histories were considered for THA and responses of the SDOF system are evaluated. The results from LabVIEW were validated and the accuracy of the algorithms generated are discussed. It is observed that the accuracy and stability of the final solution depends on the variation of displacement, velocity and acceleration that is assumed in each step. Thus, LabVIEW workbench can therefore be recognized as an effective instrument in structural engineering owing to its fast sampling features.

Keywords: LabVIEW, central difference method, wilson-θ method, SDOF system, time history analysis

1. Introduction

With the rapid development of hardware and software technology for personal computers (PCs), it is simple to effectively incorporate PCs in various precise measurement and complex control applications.VI (Virtual Instrumentation) has evolved into a thorough quest that encompasses the whole field of computer-based instrumentation leading to the large reduction of hardware. The LabVIEW (Laboratory Virtual Instrumentation Engineering Workbench) can be interfaced with several hardware, such as data acquisition cards, instrument control, and industrial automation [1]. LabVIEW is a platform and development environment for system design that focused on the framework of data flow programming. It enables the user to build programs with graphics rather than text code. It performs many applications, such as data acquisition, data interpretation, signal detection, signal

IntechOpen

processing, control and monitoring. It also simulates the vibration testing and vibration signal processing. It is an important technique that makes it easy to detect internal damage to the structure. Therefore, it is shown to be the prevailing instrument in the study of the dynamic behavior of structures which had become a major concern of mechanical, civil and aerospace engineers. To better understand the dynamic behavior, it is essential to know the modal parameters of the structure, i.e. its natural frequencies, mode shapes and damping ratios. The precise identification of these parameters can be made through the use of robust and reliable methods that belong to the field of research known as modal analysis [2].

There are different causes of vibration, such as continuous force, degradation, resonance, etc. The response of it can be understood through various control actions such as manual, automatic, sine wave generation and square wave generation on the structure. The preventive measures on the structure may be taken through analysis and monitoring of vibration signal by two processes. When the variation of force with time is known, the variation of response is formulated in time domain. This is referred to as time-domain analysis and this former signal analysis can be used to evaluate the response of any linear SDOF system to any arbitrary input. Sometimes, the force function is random and it is not possible to determine its frequency. Moreover, it may have a variable frequency over its duration and hence it is then convenient to perform the analysis in frequency domain. The frequency domain approach is also conceptually similar to the Fourier analysis procedure. However, to apply the periodic load technique to arbitrary loading, it is necessary to extend the Fourier series concept to the representation of non-periodic functions. Various researches are made in recent years to apprehend the dynamic behavior of the structure using virtual instrument engineering workbench. Sura et al., [3] analyzed the cantilever beam using the virtual instrument in which free vibrations were induced and measured in the beam. The results in the form of modal frequency were obtained for the cantilever beam which was properly fixed and he concluded that the theoretically calculated natural frequency and the experimentally calculated natural frequency are almost the same. Yao et al. [4] built a virtual earthquake simulation system instrumentation and stated that the design concept of LabVIEW is more user-friendly and efficient than others. Hu [5], describes the development of modal recognition computing tools and long-term dynamic monitoring in the LabVIEW framework. These consist mainly of two independent functional toolkits known as Structural Modal Identification (SMI) and Continuous Monitoring (CSMI) respectively. It involves checking the latest output measurements, identifying the maximum vibration amplitudes and performing statistical time series on acceleration. It generates waveform plots to represent the distribution of the frequency component and modal parameter based on automated Enhanced Frequency Domain Decomposition (EFDD) technique. An attempt is made to expand hands-on activity-based educational module through the integration of PASCO models, LabVIEW, NI hardware, sensors, and MATLAB software. Despite some existing limitations, the results successfully showed that this structure worked precisely and stably, producing good output data. It was proved as a potential tool for structural dynamics as well as Structural Health Monitoring (SHM) education and also study in which, each case of damaged structure had a distinctive property [6]. Ugo Andreaus [7] studied the experimental dynamic response of a base-isolated SDOF oscillator and formulated numerical model excited by a harmonic base acceleration using LabVIEW. The behavior of the system was well understood as the numerical simulation in LabVIEW platform efficiently agreed with the experimental investigation.

In this context of the study, an attempt is being made to propose program for time integration method in LabVIEW to predict the changes in displacement, velocity and acceleration for SDOF model for earthquake excitations. The versions of these expressions can be used for damaged structures, if the damage parameters are known. The expressions are integrated into an algorithm [8]; priory developed for Time History Analysis (THA) of structures and are analyzed here in case of central difference method and Wilson-θ method.

2. Description of the THA method

There are many numerical integration methods available to evaluate the approximate solution of equation of motions. There are two basic characteristics of these methods firstly, the differential equations of these methods are satisfied only at discrete time intervals Δt and secondly, a variation in displacement, velocity and acceleration is assumed within each time interval Δt [9]. Causevic et al., [10] discussed about non-linear dynamic time-history analysis; non-linear static method (Euro code 8); non-linear static procedure NSP (FEMA 356) and improved capacity spectrum method CSM (FEMA 440). An eight-storey reinforced concrete frame building is analyzed as the research subject. It is evident that neither of static procedures takes into consideration the damage which can be significant for long duration earthquakes. The author thus concluded that the non-linear THA was the most accurate method. Lestuzzi et al., [11] discussed about the selection of real ground motion records by considering the response of single-degree-of-freedom (SDOF) system with bilinear hysteretic model. The findings from this study are very limited, i.e., they are applicable only for building structures that can be modeled as a SDOF system. The response parameters considered are maximum displacement and ductility of the SDOF system. The study csoncludes the following points: 1.While selecting the real records of THA, the spectral acceleration records that matches with the design spectrum has to be chosen. 2. The period has to be kept as T_0 or in a range between T_0 and the period corresponding to the secant stiffness. It is observed that the mathematical computation of these methods is difficult and is time consuming and hence a requirement for alternate and efficient platform is needed.

Thus, the concept of nonlinear behavior of structures and the importance of Time history analysis (THA) is more important even though it's a century old concept. Although the linear elastic analysis and the design methods are well established, nonlinear inelastic analysis and their application to design are still evolving. The answer for the question, "Why do we need a nonlinear analysis?" lies in the fact that under extreme probable loading like earthquake; it is no longer advisable to keep the structure elastic due to the reason of yielding in structural components. Thus, a nonlinear analysis requires a clear understanding of the stress–strain curves of all the materials used in the structure, its inelastic behavior, failure criteria of the components, the capacity of its in failure modes and also the nonlinear analysis techniques. In case a single degree of freedom (SDOF) system or a multi degree of freedom (MDOF) system is subjected to a random acceleration time history, it is very difficult to solve the differential equation using the basic principle of calculus. The direct integration methods or step-by-step integration methods are used for the solutions of such problems. A very small time step Δt, is chosen and the solution is obtained from one step to the next step leading to the linear interpolation of the forces. The expression at time step (t + h) may be entirely in term of quantities at time step t or both at time step t and (t + h) which gives rise to two types of algorithm: explicit algorithm and implicit algorithm. In the former, the expressions at time step (t + h) are in terms of time step t only, whereas, in the latter, the expressions at time step (t + h) are in terms of t and (t + h). The solutions using the explicit algorithm are as easy as compared to those using the implicit algorithm.

Hence an attempt is being made to make LabVIEW programs for the widely used explicit and implicit algorithm. A brief overview of these approaches is given, followed by programming in LabVIEW platform and their validation through examples. Li [12] stated that Finite difference method optimizes the approximation for the differential operator in the central node of the considered space and provides numerical solutions to differential equations. It is noticed that the results of the central difference method approximation show a significant improvement in the accuracy along the smooth region. He also concluded that it is possible to test the function f (x) at values on the left and right of x, to obtain an optimal two-point approximation which includes abscissas that are symmetrically chosen on both sides of x. The advantage of this approach is that, its convergence speed is higher than some other finite differentiating methods, such as forward and backward differentiation. Similarly another method developed by E L Wilson for unconditionally stable linear acceleration method is Wilson θ method. This method is based on the assumption that acceleration varies linearly over an extended time step $\delta t = \theta\delta t$ [13]. Wilson-θ method is highly stable numerically as it converges rapidly to a meaningful solution. In our study, earthquake-induced ground motions of El Centro (1940) and Loma Prieta (1989) earthquake data are fed as input to the SDOF system. Seismic responses considered were in the form of acceleration, velocity, displacement and force and the application example considered was SDOF system. The accuracy which means the chosen numerical methods should converges the exact solution in terms of amplitude accuracy or amplitude decay or period accuracy or period decay was carried out in the workbench. The ground motions records were obtained from the PEER Strong Motion Database (http://peer.berke ley.edu/smcat /) [14].

3. Methodology

In the explicit method the response at time t_{n+1} is known in terms of known variables at time t_n. Thus the response values displacement, velocity and acceleration can be determined directly. Whereas, in implicit method, the response at time t_{n+1} is known in terms of the known variables at time t_n and unknown variables at time t_{n+1}. These implicit algorithms involve either an iterative scheme or solution of linear simultaneous equations because the unknown quantities appear on both sides of the equations.

3.1 Central difference method

This method is based on the finite difference approximation of the time derivative of displacement, that is, velocity and acceleration [9].

An equation of motion for an SDOF system is given as:

$$m.\ddot{u} + c.\dot{u} + k.u = F_t \tag{1}$$

m = Mass, c = Damping, k = Stiffness, $\ddot{u}$ = Acceleration, $\dot{u}$ = Velocity, u = Displacement, F_t = Force.

Initial acceleration is given as,

$$\ddot{u}_o = \frac{-m.\Delta\ddot{u}_g - c.\dot{u}_o - k.u_o}{m} \tag{2}$$

Initial displacement at i-1th time step

$$u_{i-1} = u_0 + \Delta t(\dot{u}_o) + \frac{(\Delta t)^2}{2}\ddot{u}_o \tag{3}$$

Incremental Stiffness for the ith time step,

$$k_i = \frac{m}{(\Delta t)^2} + \frac{c}{2.\Delta t} \tag{4}$$

Incremental force for ith time step,

$$F_i = -m.\ddot{u}_g - a.u_{i-1} - b.u_i \tag{5}$$

Where, *a* and *b* are constants and given as

$$a = \frac{m}{2.\beta} + \frac{\gamma.c}{\beta}, b = \frac{m}{2.\beta} + \Delta t.\left(\frac{\gamma}{2.\beta} - 1\right)c \tag{6}$$

Displacement for i + 1th time step,

$$u_{i+1} = \frac{F_i}{k_i} \tag{7}$$

Velocity for ith time step,

$$\dot{u}_i = \frac{u_{i+1} - u_{i-1}}{2\Delta t} \tag{8}$$

Acceleration for ith time step,

$$\ddot{u} = \frac{u_{i+1} - 2.u_i - u_{i-1}}{(\Delta t)^2} \tag{9}$$

3.2 Algorithm

Step 1: Initial displacement and velocity are known as initial conditions of the problem at time t = 0.
Step 2: Damping c and stiffness k are computed from the system properties.
Step 3: Acceleration at time t = 0 is computed from Eq. (2).
Step 4: Compute equivalent stiffness $k_i$from Eq. (4).
Step 5: For time step i, compute equivalent force F_i from Eq. (5).
Step 6: Compute constants a and b from Eq. (6).
Step 7: Solve for new displacement u_{i+1} from Eq. (7).
Step 8: Compute velocity and acceleration at time step i from Eq. (2) and (3).
Step 9: Repeat Steps 6 to 8 for the next time step.

3.3 Programs developed in LabVIEW for central difference method

A visual block diagram which describes the data flow within the VI is presented in the form algorithm in LabVIEW. LabVIEW accepted the input, and the algorithm was sampled and programmed through appropriate interfaces in accordance with the specification of VI, and the output data was collected. In our software, data such as damping, mass and time period were provided as an input and displacement, velocity and acceleration plot was obtained for the time history data (i.e. it can function as an analog to digital converter). Owing to the sheer quantity and simplicity of the different built-in functions, the data was thus manipulated in a wide range of forms as shown in **Figure 1**.

3.4 Wilson-θ method, linear SDOF system

The incremental equation of equilibrium known as Wilson-θ method is developed by Prof.E.L.Wilson, University of California, and Berkeley [9]. The calculations are carried out over an extended time step θΔt, where θ is an amplifier for the time step. It assumes that the variation of acceleration over the extended time step remains unchanged, that is, it is still the same as that of the original time step Δt, a linear variation.

An equation of motion for an SDOF system is given as:

$$m.\ddot{u} + c.\dot{u} + k.u = F_t \tag{10}$$

m = Mass, c = Damping, k = Stiffness, $\ddot{u}$ = Acceleration, $\dot{u}$ = Velocity, u = Displacement, F_t = Force.

Initial acceleration is given as,

$$\ddot{u}_o = \frac{-m.\Delta\ddot{u}_g - c.\dot{u}_o - k.u_o}{m} \tag{11}$$

Incremental force for i^{th} time step,

$$\delta F_i = \theta\left(-m.\Delta\ddot{u}_g\right) + a.\dot{u}_i + b.\ddot{u}_i \tag{12}$$

Where, a and b are constants and given as

$$a = \frac{6.m}{q.\Delta t} + 3c, b = \frac{q.\Delta t.c}{2} + 3.m$$

Tangent Stiffness for i^{th} time step,

$$k_t = k_i + \frac{3.c}{\theta.\Delta t} + \frac{6.m}{\theta(\Delta t)^2} \tag{13}$$

Solve for δu for i^{th} time step,

$$\delta u_i = \frac{\delta F_i}{k_t} \tag{14}$$

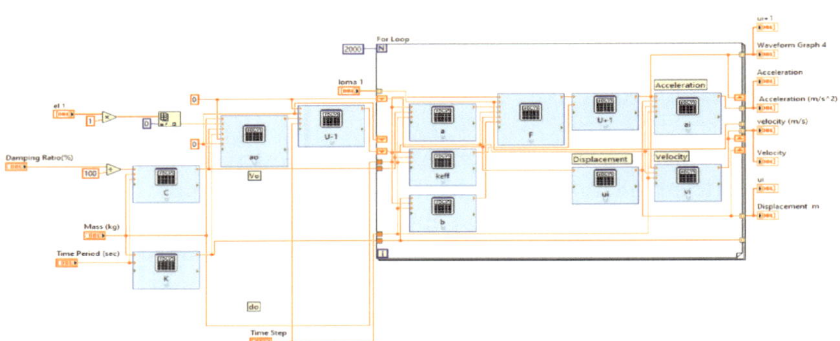

Figure 1.
Block diagram of central difference method in LabVIEW.

Solve for $\delta\ddot{u}$ for ith time step,

$$\delta\ddot{u}_i = \frac{6.\delta u_i}{(\theta.\Delta t)^2} - \frac{6.\dot{u}}{\theta.\Delta t} - 3\ddot{u}_i \tag{15}$$

Incremental Acceleration is given as

$$\ddot{u}_i = \frac{\delta\ddot{u}_i}{k_t} \tag{16}$$

Knowing incremental acceleration, incremental velocity and displacement can be calculated,

Incremental velocity,

$$\Delta\dot{u}_i = \Delta t.\ddot{u}_i + \frac{\Delta t.\Delta\ddot{u}_i}{2} \tag{17}$$

Incremental displacement,

$$\Delta u_i = \Delta t.\ddot{u}_i + \frac{(\Delta t)^2\dot{u}_i}{2} + \frac{(\Delta t)^2\Delta\ddot{u}_i}{6} \tag{18}$$

At time t_{i+1} displacement, velocity and acceleration can be calculated as

$$\begin{aligned} u_{i+1} &= u_i + \Delta u_i \\ \dot{u}_{i+1} &= \dot{u}_i + \Delta\dot{u}_i \\ \ddot{u}_{i+1} &= \ddot{u}_i + \Delta\ddot{u}_i \end{aligned} \tag{19}$$

3.5 Algorithm

Step 1: Compute k_t.

Step 2: Calculate u_{i+1},$\dot{u}_{i+1}$, $\ddot{u}_{i+1}$ using the Eqs. (15), (17) and (18).

Step3: Update c and k.

Step 4: Repeat steps 1 to 3.

It should be noted that in this method, k and c are assumed to remain constant during the extended time step and are updated at the end of the real-time increment **Δt.θ** = **1** leads to the linear acceleration method. It is recommended that **θ** is taken >1.37 [15].

3.6 Programs developed in LabVIEW

The block diagram given below shows all the features that are expressed in VIs. The input signal was simulated at the first step. This was accomplished by

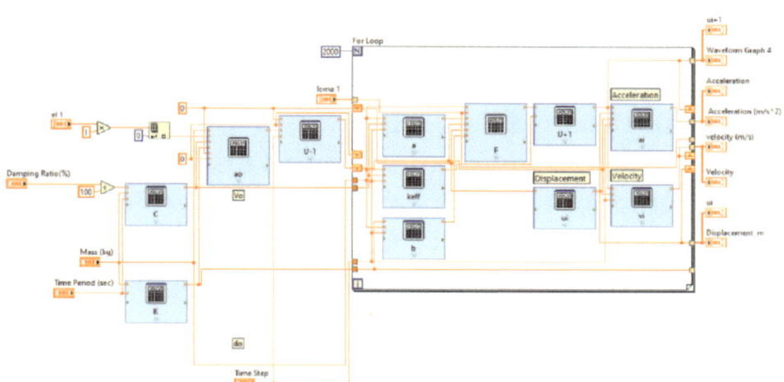

Figure 2.
Block diagram of wilson-θ method in LabVIEW.

incorporating block diagram from simulate signal express VI, which is located under the signal analysis palette. The functions, such as mathematical operations, express VIs, built analysis tools and signal simulation, were assessed from the palettes by right clicking in the block diagram or front panel, which brought up the palette menu. The frame work was thus created using the algorithm and the waveform was generated for displacement, acceleration and velocity as shown in the **Figure 2**.

4. Analytical validations

The peak ground motion recorded, magnitude and it's predominated period at real-time data storage station during the 1940 El Centro earthquake or 1940 Imperial Valley earthquake (Mw = 6.9) was considered for THA as the first analytical case study. The SDOF system that was considered has a mass of 1 kg and a damping value of 0.05. The time step that was considered for it was 0.02 s. The ground motion details (horizontal component) are given in **Table 1**. In order to further prove the efficiency of the program, the 1989 Loma Prieta earthquake was considered for analysis and the percentage variation of the LabVIEW is evaluated. The basic parameters of the SDOF system considered has a mass of 1 kg, time step 0.02 s, damping ratio 0.05 and time period of 0.513 s [16].

Earthquake	Maximum acceleration (g)	Magnitude	Predominant period
El Centro	0.296	6.9	0.588
Loma Prieta	0.276	7	0.588

Table 1.
Earthquake ground motion details (horizontal component).

5. Results and discussion

5.1 Linear SDOF system response of time integration methods and its results and discussion

El Centro and Loma Prieta earthquake ground motions were considered for the analysis. The problem was solved using the time step 0.02 to understand the displacement-time history, velocity-time history and acceleration- time history under El Centro and Loma Prieta earthquake using both the methods on LabVIEW and are shown below.

5.1.1 Response in terms of displacement of linear SDOF system

The displacement response in central difference and wilson-θ method were obtained in LabVIEW and was displayed below in **Figures 3** and **4**.

In case of El Centro earthquake maximum peak displacement of 0.0531 (m) and minimum peak displacement of 0.0456 (m) was given by Wilson-θ method and in case of Loma Prieta the maximum peak displacement of 0.033 (m) and minimum peak displacement of 0.029 (m) was again given by Wilson θ-method.

5.1.2 Response in terms of velocity of linear SDOF system

The velocity response in central difference and wilson-θ method are obtained in LabVIEW and is displayed below (**Figures 5** and **6**).

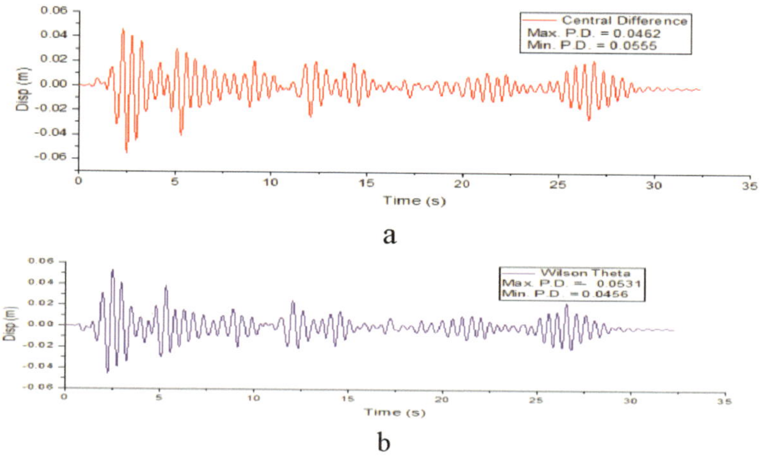

Figure 3
Displacement vs time response under El Centro earthquake (a) central difference method (b) wilson-θ method.

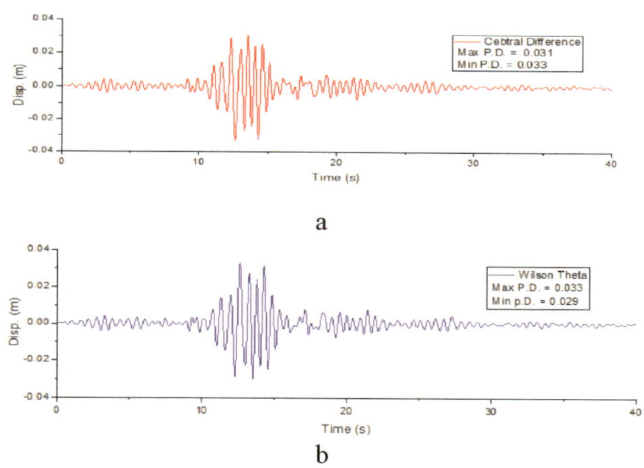

Figure 4.
Displacement vs time response under Loma Prieta earthquake (a) central difference method (b) wilson-θ method.

The complex solution determined by the central difference method in terms of velocity is contrasted with the method of wilson-θ. In the case of El Centro earthquake highest peak velocity of 0.653 (m/s.) and lowest peak velocity of 0.606 (m/s.) is responded by wilson-θ method and in case of Loma Prieta the highest peak velocity of 0.363 (m/s.) and lowest peak velocity of 0.341 (m/s.) is displayed by central difference method.

5.1.3 Response in terms of acceleration of linear SDOF system

The acceleration response in central difference and wilson-θ method were obtained in LabVIEW and are displayed below (**Figures 7** and **8**).

Dynamic acceleration response calculated using central difference was compared with wilson-θ method. In case of El Centro earthquake motion, maximum peak acceleration of 10.32 (m/s^2) and minimum peak acceleration of 7.75 (m/s^2) was displayed by central difference method. Whereas, in case of Loma Prieta

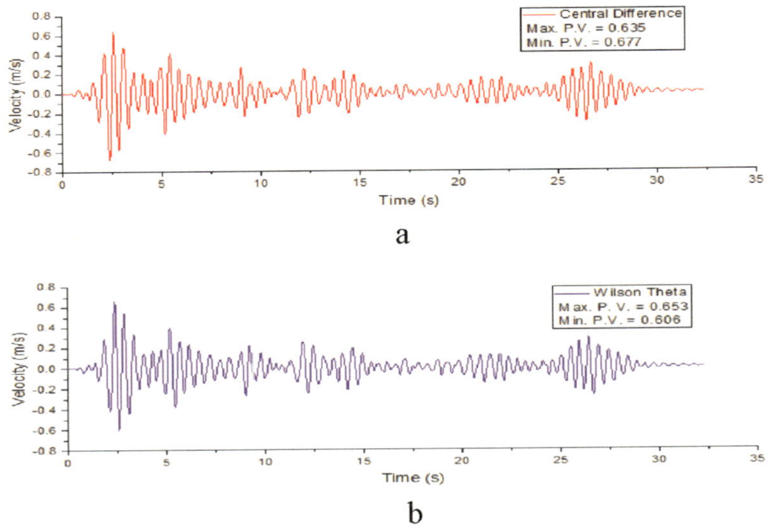

Figure 5.
Velocity vs time response under El Centro earthquake (a) central difference method (b) wilson-θ method.

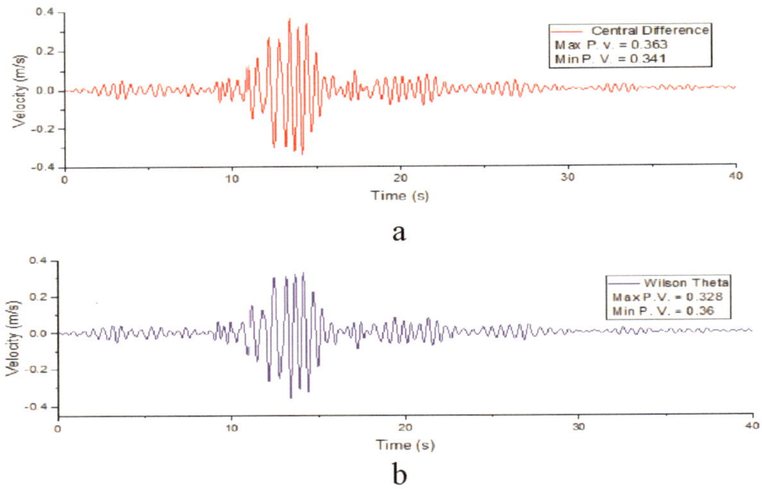

Figure 6.
Velocity vs time response under Loma Prieta earthquake (a) central difference method (b) wilson-θ method.

ground motion the maximum peak acceleration of 4.25 (m/s^2) was shown by central difference method and minimum peak acceleration of 4.24 (m/s^2) was given by wilson-θ method.

In the **Table 2** shown above, response results obtained from central difference method and wilson-θ method for El Centro earthquake was worked out. The difference in response was calculated and it was clearly seen that the variation did not exceed more than 0.08.

In the **Table 3** shown above, response results obtained from central difference method and wilson-θ method for Loma Prieta earthquake was worked out. The difference in response was calculated and it was clearly seen that the variation did not exceed more than 0.04.

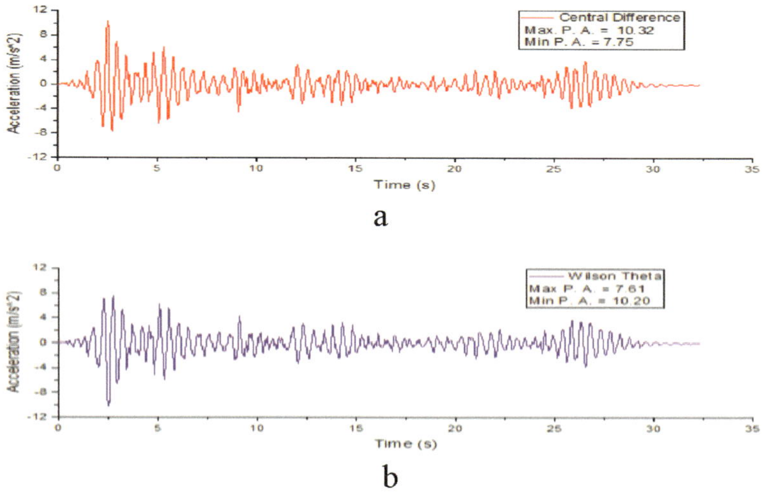

Figure7.
Acceleration vs time response under El Centro ground motion (a) central difference method (b) wilson-θ method.

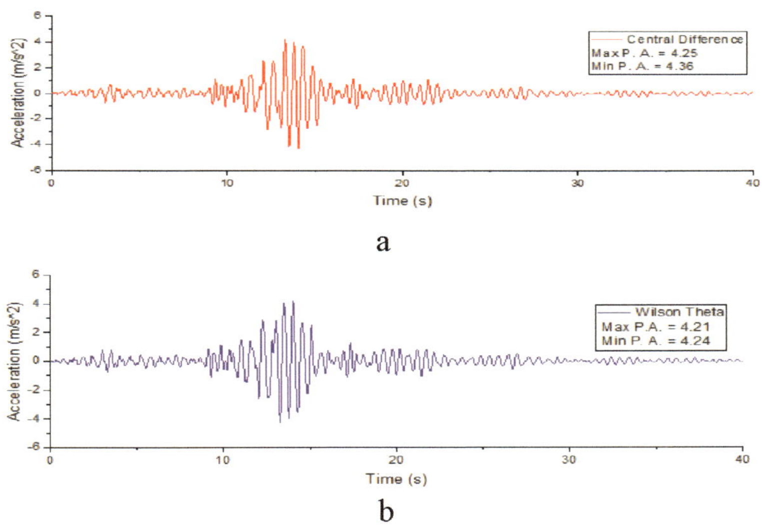

Figure 8.
Acceleration vs time response under Loma Prieta ground motion (a) central difference method (b) wilson-θ method.

El Centro	Peak Displacement(m)	Peak Velocity (m/s)	Peak acceleration (m/s^2)
CDM	0.0462	0.635	10.32
WTM	0.0531	0.653	10.20

Table 2.
Percentage variation of linear SDOF system under El Centro earthquake.

6. Conclusions

This paper summarizes, the modeling of linear SDOF system in LabVIEW software using time integration method. The comparative study with the results of an example chosen for the proposed program in LabVIEW clearly stated that the

Loma Prieta	Peak Displacement(m)	Peak Velocity (m/s)	Peak acceleration (m/s^2)
CDM	0.031	0.363	4.25
WTM	0.033	0.328	4.21

Table 3.
Percentage variation of linear SDOF under Loma Prieta earthquake.

responses obtained are accurate and hence programming in it for time integration problems will lead to trustworthy results.

These methods are thus based on two essential features:

1. Variation of displacement, velocity and acceleration are assumed within each time step. Hence, the accuracy and stability of the final solution depends on this variation.

2. The equilibrium is satisfied at all discrete time points within the interval of solution instead of any time t.

As the percentage difference between central difference method and wilson-θ method is negligible these programs can be extended to various earthquake ground motions and also for non-linear simulations. There is a need to track the system displacement, member spring force and spring stiffness to solve the problems in material nonlinearity. It should be noted that the time steps and appropriate stiffness has to be chosen very carefully to obtain accurate results. With the establishment of appropriate modeling for various other integration methods the accuracy can be further improved and time constraint problems can be easily solved.

Author details

R.B. Malathy*, Govardhan Bhat and U.K. Dewangan
Department of Civil Engineering, NIT Raipur, Raipur, India

*Address all correspondence to: malathyrbr@gmail.com

References

[1] K. Mahant and C. Bhatt, "Design and Development of Detector Simulator for Total Ionized Dose and ground checkout system of radiation monitoring instrument," Int. J. Electron. Telecommun., vol. 63, no. 4, pp. 431–436, 2017, doi: 10.1515/eletel-2017-0059.

[2] I. C. Mituletu, G. R. Gillich, and N. M. M. Maia, "A method for an accurate estimation of natural frequencies using swept-sine acoustic excitation," Mech. Syst. Signal Process., vol. 116, pp. 693–709, 2019, doi: 10.1016/j.ymssp.2018.07.018.

[3] S. Sura, A. Sawale, and M. S. Gupta, "Dynamic analysis of cantilever beam," Int. J. Mech. Eng. Technol., vol. 8, no. 5, pp. 1167–1173, 2017.

[4] K. C. Yao, W. T. Huang, C. L. Lin, P. E. Wu, and J. S. Chiang, "Virtual Instrumentation Design on Earthquake Simulation System," no. Aiie, pp. 609–612, 2015, doi: 10.2991/aiie-15.2015.162.

[5] W. H. Hu, Á. Cunha, E. Caetano, F. Magalhães, and C. Moutinho, "LabVIEW toolkits for output-only modal identification and long-term dynamic structural monitoring," Struct. Infrastruct. Eng., vol. 6, no. 5, pp. 557–574, 2010, doi: 10.1080/15732470903068672.

[6] Tu Hoang, "Development of dynamic simulators using Pasco models equipped with LabVIEW," pp. 1–35.

[7] A. Korgin, V. Ermakov, and L. Z. Kilani, "Automation and Processing Test Data with LabVIEW Software," *IOP Conf. Ser. Mater. Sci. Eng.*, vol. 661, no. 1, 2019, doi: 10.1088/1757-899X/661/1/012073.

[8] G. R. Gillich, H. Furdui, M. Abdel Wahab, and Z. I. Korka, "A robust damage detection method based on multi-modal analysis in variable temperature conditions," Mech. Syst. Signal Process., vol. 115, pp. 361–379, 2019, doi: 10.1016/j.ymssp.2018.05.037.

[9] S. Rajasekaran, *Structural Dynamics of Earthquake Engineering; Theory and Application using Mathematica and MATLAB*, vol. 9781439801. 2009.

[10] M. Causevic and S. Mitrovic, "Comparison between non-linear dynamic and static seismic analysis of structures according to European and US provisions," Bull. Earthq. Eng., vol. 9, no. 2, pp. 467–489, 2011, doi: 10.1007/s10518-010-9199-1.

[11] P. Lestuzzi, Y. Belmouden, and M. Trueb, "Non-linear seismic behavior of structures with limited hysteretic energy dissipation capacity," Bull. Earthq. Eng., vol. 5, no. 4, pp. 549–569, 2007, doi: 10.1007/s10518-007-9050-5.

[12] J. Li, "Computational Fluid Dynamics.," *Cambridge Univ. Press. 2002. 1012 pp. ISBN 0 521 59416 2. J. Fluid Mech. 491, 411–412. doi10.1017/S0022112003005445*.

[13] S. Dhakal, N. P. Bhandary, R. Yatabe, P. L. Pradhan, and R. C. Tiwari, "Finite Element Modeling and Decoupled Seismic Stability Analysis of a Zoned Rockfill Dam Designed By Traditional Empirical Methods," J. Inst. Eng., vol. 8, no. 1–2, pp. 71–92, 1970, doi: 10.3126/jie.v8i1-2.5098.

[14] "Pacific Earthquake Engineering Research (PEER) Center (2013), Ground Motion Database. Available from: http://peer.berkeley.edu/peer_ground_motion_database [20 July 2013]."

[15] D. 10. 1007/97.-1–4615–0481-8 Paz, Mario. Structural Dynamics: Theory and Computation I -5th ed. p.cm.ISBN 978–1–4613-5098-9 and Includes, *STRUCTURAL*. .

[16] R. B. Malathy, G. Bhat, and U. K. Dewangan, "Generalized logic for modeling and obtaining harmonics estimation in shake table using LabVIEW with experimental validation," Asian J. Civ. Eng., vol. 21, no. 6, pp. 985–994, 2020, doi: 10.1007/s42107-020-00255-x.

Chapter 8

LabVIEW as Power Disturbances Classification Tools

Ahmad Farid Abidin and Mohd Abdul Talib Mat Yusoh

Abstract

Power disturbances monitoring is one of the important aspects on dealing power quality issue in electrical system. The aims of conducting monitoring process are to identify the real culprit which contribute to the Power Quality (PQ) problem. One of the vital steps during monitoring process is the classifying various type of power disturbance. This classification process is very important to give a right direction towards proposing the correct mitigation technique. In order to produce reliable classification technique, the devices which has a flexibility on accommodating the software and hardware part need to be deployed. The software is need for algorithm development such as signal processing, Artificial Intelligent (AI) as well as statistical analysis. On the hardware part, the device's ability to acquire the electrical parameter within the electrical system operation is very important. The data acquisition based on the voltage and current is essential to be feed in the classification algorithm in software side. On the other hand, the interfacing devices and data acquisition module need to be developed at the hardware side, LabVIEW manage to accommodate both software and hardware need and further development of the LabVIEW for this purpose will be elaborated in this chapter.

Keywords: LabVIEW, power quality (PQ)

1. Introduction

Power quality is a term that commonly used to represent the electrical supply to the consumer without any disruption or blackout. Technically, power quality is defined as the availability of supply voltage to be in the sinusoidal form within the permissible magnitude and frequency without misoperating the electrical equipment [1, 2]. The power quality issue becoming more important since 1980's due to the dominant use of electronic equipment among the electrical consumers. The electronic equipment is sensitive to any deviation in term of wave shape, magnitude, and frequency of supply voltage as their performance could be affected i.e. misoperation, burn, or shorten lifespan. Nowadays, most of the industries and commercial electrical consumer used the electronic equipment in their daily operation. Hence, the issues of power quality are very prevalent in business environment as the industry and commercial electrical consumer complained that they experienced power quality problem regularly.

Once the electric supply has a low power quality, all the undesirable situation would be experienced by electrical user. In order to have a good power quality, the electrical supply need to have a voltage with the attribute of pure sinusoidal(5% deviation), the permissible magnitude (nominal voltage ±10%) and 50 Hz frequency with permissible range(±1%) [3, 4]. Based those three attributes, any deviation from the given permissible limit is known as power disturbances. The power disturbances consist of different type of disturbances namely, voltages sags, swell, interruptions, transient over voltages, harmonics, and voltage imbalance. The definitions of each type power disturbances can be described as follow [5]:

Referring to **Table 1**, the type of disturbances has its own unique characteristics. Therefore, there are a paramount important to classify the type disturbance at initial stage of the PQ solution step. The classification of power disturbance is occasionally incorporating with signal processing technique which functioning as feature extraction tools. The popular signal processing technique that often used in extracting the feature are Short time Fourier Transform (STFT), Wavelet Transform (WT), S-Transform (ST), Hilbert Transform (HT), Hilbert Huang Transform (HHT), and Ensemble Empirical mode Decomposition (EEMD) [6–11]. STFT had been widely used as a tool in recognizing disturbances. Nevertheless, STFT has a limitation where it's only has a fixed windowing technique which is not fit for non-stationary signal [12]. The improved WT has been introduced to enable the nonstationary signal processing approach in continuous windows size by using multiresolution analysis (MRA) [13]. However, this technique suffers from high losses of information during MRA process [14]. On top of that, the WT has disadvantage where it fails to extract time-frequency information.

S-transform (ST) is proposed to carry out the solution which experienced by STFT and WT [15, 16]. ST is superior signal processing tool, where its technique is improved and outcome the limitation that exist in STFT and WT. ST manage to provide the distinct features which facilitate the decision making tools i.e., Artificial Intelligent (AI) to produce the accurate classification result accuracy. From the literature studies, there are several types of AI that frequently used as a decision making mechanism for classifying power disturbances, such as the Support Vector Machine (SVM), Levenberg Marquardt Neural Network (LMNN), Probability Neural Network (PNN), General Regression Neural Network (GRNN), and Radial Basic Function Neural Network (RBFNN) [17–21]. However, due to the less computational burden and resilience performance, the SVM has been chosen as a decision-making tool to be used in this work. Both ST and SVM are embedded to

Type of power disturbance	Characteristic
Voltage sag	Temporary voltage reduction in magnitude between 10% and 90% of nominal voltage in RMS for duration few cycles to one minute
Interruptions	Interruptions can be defined as 0.9 pu reductions in voltage magnitude for duration less than 1 minute.
Swell	An increase of RMS voltage from 0.1 pu to 0.8 pu for duration 0.5 cycle to 1 minute.
Harmonics	A Sinusoidal voltages or currents which having frequencies that are integer multiple of the frequency at which supply system is designed to operate.
Voltage imbalance	The ratio of the negative sequence component to the positive sequence component which exceed 0.5%
Interruptions	The disappearance of the supply voltage on one or more phases. It is usually qualified by an additional term indicating the duration of the interruption
Transient over voltages	Sudden, non-power frequency change in the steady-state condition of voltage, current, or both,

Table 1.
The definition of each type power disturbance.

LabVIEW module by means of MATLAB algorithm to perform as power disturbance classifier.

LabVIEW is fast and widely used hardware integration which is a convenient instrument and PC constructed with data acquisition that has the ability to various interfacing type of real-time input signal [22]. Thus, it offers valuable analysis with variation choice of programming tools which is suitable to be implemented in real-time power quality monitoring system.

2. LabVIEW role

This research is conducted in LabVIEW software with the aid of National Instrument (NI) data acquisition module, where the entire data acquisition module, analysis module, features extraction process and data representation are developed and executed using this tools [23]. Along with enormous flexibility in interfacing and working with various type of real time input signal from different type of measurement tools, LabVIEW is offering a superiority analysis with a variety choice of programming tools that is appropriate to be implemented in real time power quality monitoring system. The flexibility of LabVIEW software also grants an excellent output result of the projects with lesser computational time. This develops system from this research basically is able to extract features and classify the PQ disturbance as illustrated in **Figure 1**.

Based on **Figure 1**, the developed system starts with the accumulating of PQ disturbances by using the PQ recorder i.e. Fluke 1750. Then those PQ disturbances are transferred to Chroma 61701 for storing and regenerating purpose. Then a National instruments Compaq Data Acquisition Chassis (NI cDAQ-9184) with NI 9225 card is used for high-speed acquisition of PQ disturbances. Further, the types of PQ disturbances are analyzed in LabVIEW interfacing since the voltage signals

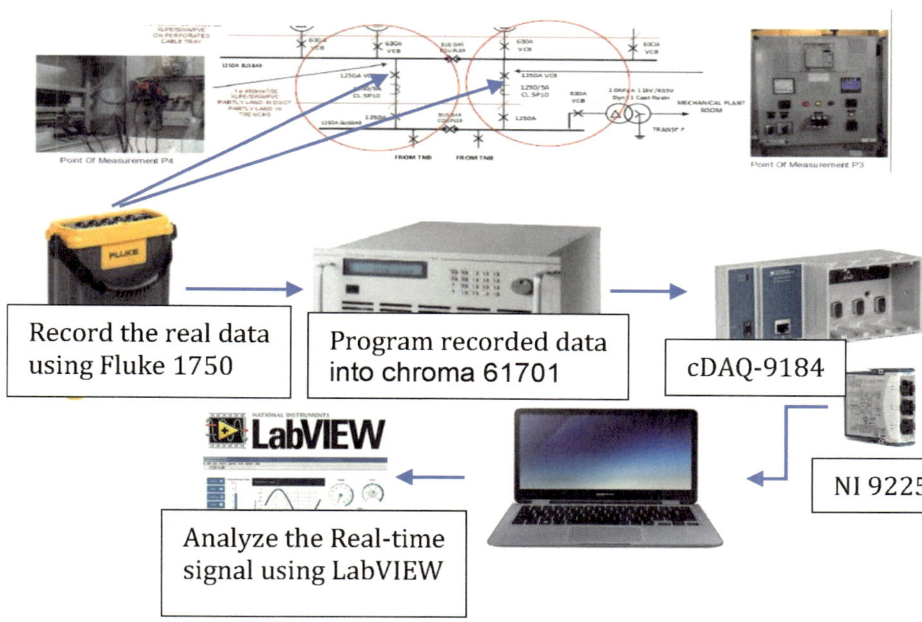

Figure 1.
System architecture of PQ analysis using LabVIEW software.

which acquired from the NI 9225 card. Further work of classification process which based on ST and SVM are developed at LabVIEW platform.

3. S-transform algorithm

Time-frequency transform as known as S-Transform was proposed by Stockwell and his co-workers in 1996, inheriting the advantage of Continuous Wavelet Transform (CWT). As a modification from CWT, ST can be derived by multiplying the phase factor to CWT equation as expressed in (1) where the algorithm of CWT can be defined as (2)

$$ST(\tau,f) = CWT(\tau,d)e^{i2\pi ft}dt \tag{1}$$

$$CWT(\tau,d) = \sum_{-\infty}^{+\infty} x(t)\psi(t-\tau,d)dt \tag{2}$$

By substituting (2) into (1), the final equation for the ST can be define as,

$$ST(\tau,f) = \sum_{-\infty}^{+\infty} x(t)\psi(t-\tau,d)e^{i2\pi ft}dt \tag{3}$$

Unlike CWT, ST applies Gaussian window instead of mother wavelet which can be defined as,

$$\psi(t,f) = g(t)e^{-i2\pi ft} \tag{4}$$

Where g(t) can be expressed as,

$$g(t) = \frac{1}{\sigma\sqrt{2\pi}}e^{-t^2/2\sigma^2} \tag{5}$$

The capability of the ST Gaussian window to varies according to the frequency of the signal is depend on the Gaussian window width, σ which can be define as,

$$\sigma(f) = \frac{1}{|f|} \tag{6}$$

Hence, the final equation of the ST can be expressed as,

$$ST(t,f) = \frac{f}{\sqrt{2\pi}}\sum_{-\infty}^{+\infty} x(t)e^{-(t-\tau)^2 f^2/2}e^{i2\pi ft}dt \tag{7}$$

3.1 S-transform embedded in LabVIEW

In order to apply signal processing analysis, the ST algorithm function is embedded in LabVIEW software via MathScript Node that act as a native compiler of m file which can blend textual and graphical approaches for signal processing and data analysis task. MathScript Node works in integrating m files into the LabVIEW graphical environment. Embedded process of m file ST algorithm into LabVIEW present a hybrid approach between text-based maths with graphical programming, which is enhanced the capability of previously developed algorithm to perform the analysis in a real time environment.

MathScript Node is an embedded LabVIEW feature that links the text based I/O variables with the inputs and outputs of LabVIEW. **Figure 2** shows ST analysis implementation in LabVIEW via MathScript Node. In order to call the user-define function from m file into the LabVIEW MathScript Node, the path for the m file directory will be added to MathScript: Search Path in MathScript Window. The input variable on the right-hand side of the MathScript Node will be the voltage signal and the output on the left-hand side of the MathScript Node indicates all possible output of the transform voltage signal by ST.

The output of transformed voltage signal using ST is in the form of N × M matrix called S-matrix. Each element within the 2-dimension array of the S-matrix is a complex value where the row represents the frequency while the column represents time. Hence, the output of the S-matrix could be visually analyzed by the representation of time-amplitude graph and frequency-amplitude graph, where the amplitude stands for the absolute value or vector of each complex element from the S-matrix. The example of one of the PQ disturbance is illustrated in **Figure 3**. Further analysis is performed on the S-matrix output in the form of time-amplitude and frequency-amplitude graph as shown in **Figures 4** and **5** respectively in order to extract the feature of PQ disturbance.

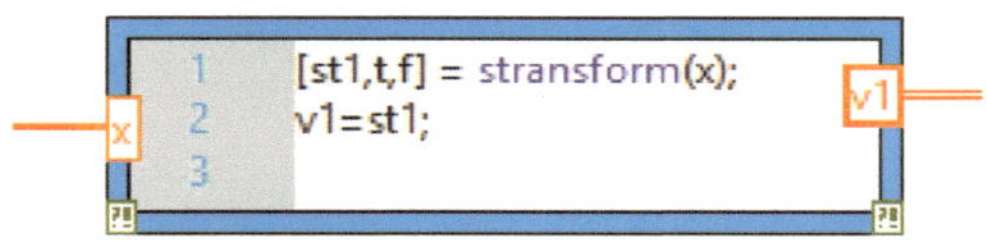

Figure 2.
LabVIEW MathScript node with user define function called S-transform output features.

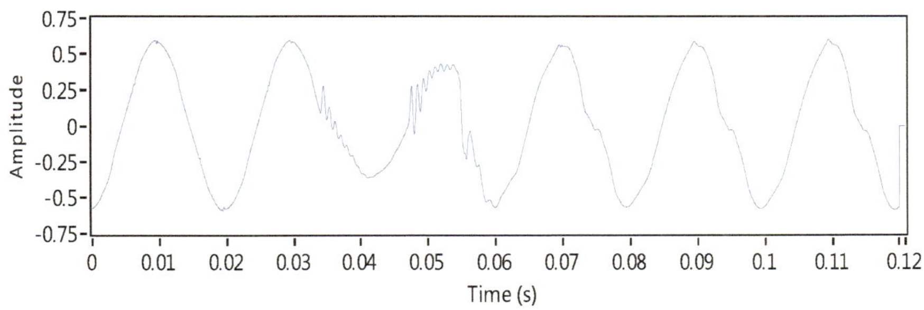

Figure 3.
IEEE working group recorded waveform of PQ disturbance.

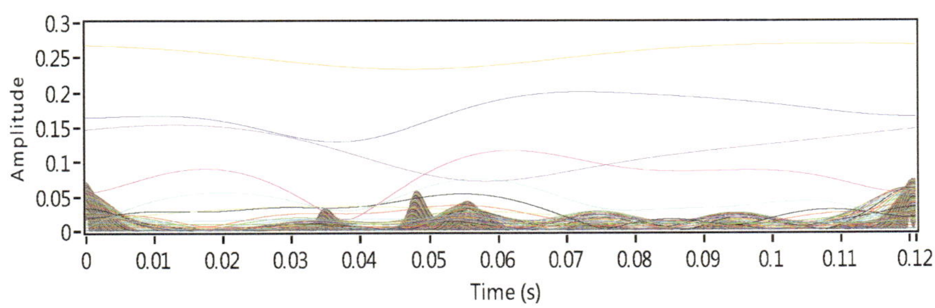

Figure 4.
Time-amplitude graph of S-matrix.

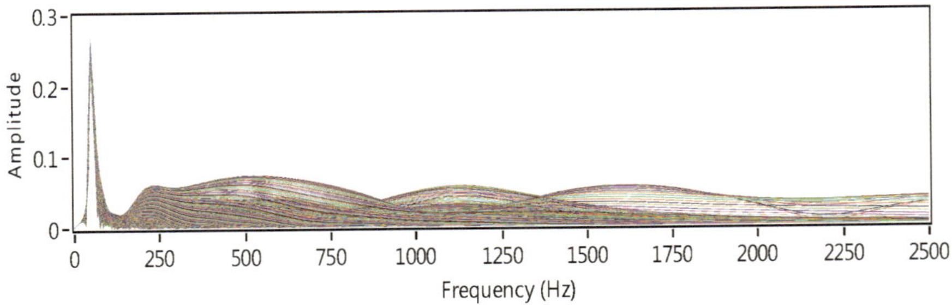

Figure 5.
Frequency-amplitude graph of S-matrix.

3.2 Support vector machine

SVM is basically a binary two-class classifier and has been extended to multiclass classifier. Multi error-correcting output code (ECOC) technique has been used to deal with multiclass classifications problems [24–26]. The SVM equation for function equation can be written as:

$$f(y) = \sum_{k=1}^{N} a_i K(x, x_i) + b \tag{8}$$

Where a_i the weighting factor, x is the training samples and x_i are the support vectors, b represent the bias and N is the training samples. In a classifier model, only the support vector takes effect in the classification. The kernel function helps nonlinear SVM maps the training sample from the input space into the higher dimensional feature space [27, 28]. Kernel function can be written as:

$$K(x, x_i) = \phi(x).\phi(x_i) \tag{9}$$

Eq. (10) below is the example of nonlinear equation for radial basis function (RBF) SVM.

$$K(x, x_i) = \exp\left(-\frac{\|x - x_i\|^2}{2\sigma^2}\right) \tag{10}$$

Where σ is the kernel width. The only difference of nonlinear SVM function after the implementation of kernel function would be:

$$f(y) = \sum_{k=1}^{N} a_i \phi(x).\phi(x_i) + b = \sum_{k=1}^{N} a_i K(x, x_i) + b \tag{11}$$

There are two types of coding design in ECOC-SVM to construct the classifiers. Most common used of coding design are One-versus-One (OVO) and One-versus-All (OVA) [26]. ECOC is used to analyze and correct the error data when conducted in the channel. The ECOC algorithm capable to enhance the performance of fault tolerance of classification model by encoding different class. The classification problem can be overcome by associating a row of a $k \; x \; m$ "coding matrix" with entries {−1, 0, +1). Each class is assigned a binary string of length m where m is the number of binary classification problems to be constructed. The comparison between classes "-1" and "+1" and ignoring classes with "0" is represented from matrix column. The "codeword" string which are $"I_t^+"$ and $"I_t^-"$ respectively used to replace "-1" and "+1". Then m binary function is learned, one for each bit position in these binary strings. The desired outputs of m binary functions are specified by the codeword of class i during the training of class i. Now, new values of X are classified to generate an m-bit strings by evaluating each of the n binary function. X is assigned to the class whose codeword is closest after comparing string to each of the k codeword [29].

LabVIEW is a tool to design a system of engineering application that capable to evaluate and control. LabVIEW is fast and widely used hardware integration which is convenient instrument and PC constructed with data acquisition that has the ability to interfacing various type of real time input signal [22].

This software holds the three components which are the front panel, block diagram, and icon and connector panel. The front panel on **Figure 6** serves as the user interface. The block diagram contains the graphical source code that defines the functionality of the block diagram. Voltage continuous block diagram based on example finder is select to capture the real time waveform. It determines a block diagram on continuously acquire voltage measurement using a DAQmx device referring on **Figure 7**. During the generation of PQ disturbance, voltage measurement NI is already connected via Ethernet to the computer to capture the real-time voltage signal. In order to gain voltage waveform, the physical channel must be set to ai1. To gain accurate waveform the sample rate and number of samples are set to 12800 and 256, respectively.

The block diagram of voltage continues to capture and extract all the features of voltage signal with the aid of ST algorithm. While loop is constructed to block diagram at acquired data in order to repeats the code within its sub diagram until a specific condition occurs. A While Loop always executes at least one time. The mathscript node is built to insert the ST algorithm. The input variable from the mathscript node is connect to DAQmx in order to read one or more waveforms from a task that contains one analog input channels which is voltage measurement NI. Output variable from the mathscript node is inserted into array block + feedback

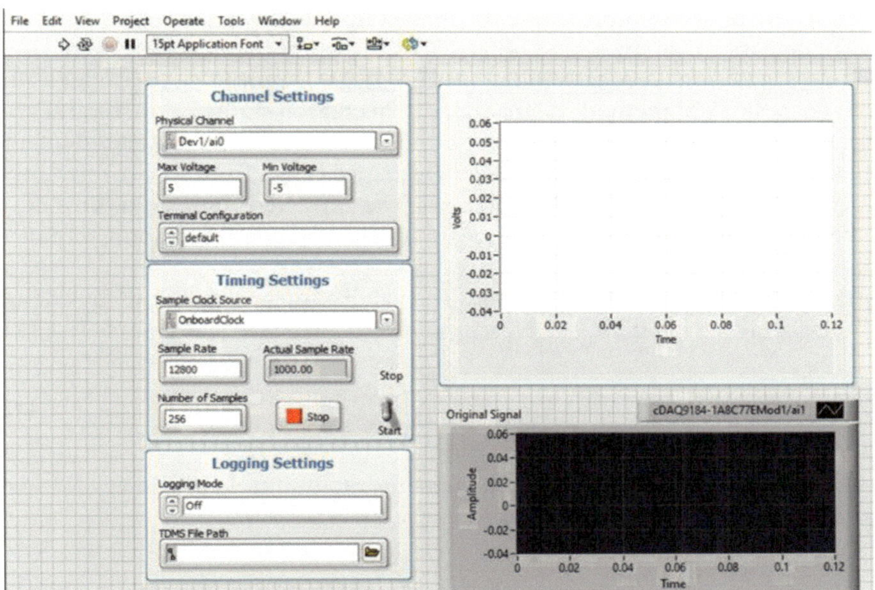

Figure 6.
Front panel of LabVIEW software.

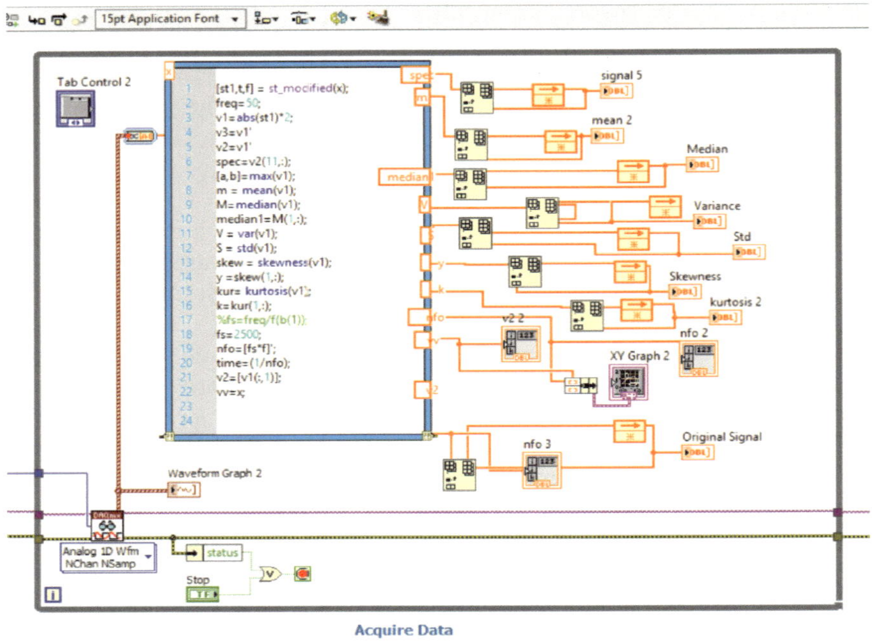

Figure 7.
Block diagram of high NTGV features extraction.

and then connect to the waveform graph. The feedback on the array block is utilized to store and display each cycle of data per windows on the waveform graph. This is a simple manner to capture the features of PQ problems according to real-time signal of measurement. It is also to gain S-matrix graph of frequency time represent time series signal. Inside this mathscript node, MATLAB coding of ST and statistical command are embedded.

Extracted features will be in statistical data waveform. All the wanted features extraction coding is assemble inside the mathscript node which are mean, median, standard deviation, variance, and others. This statistical method suitable to be used when lot of data is picked for training process and it is profitable to produce an effectiveness in SVM method of classifying. This reason statistical method is choosing because the data collected is in a set of possible value of the measured quantity.

All the extracted features of the PQ disturbance signal from ST are introduce in the form of the frequency-amplitude graph. The extracted features in form of statistical data are perform based on mean, median, standard variation, variance, kurtosis and skewness.

Mean is the average value of a signal. The value of mean can be gain by adding all amplitude of signal and divide by N, sample. Mathematical equation as below (12).

$$\mu = \frac{1}{N}\sum_{i=1}^{N-1} x_i \tag{12}$$

Median is a middle number of arrangements of amplitude in numerical order and the mathematical equation state below:

$$m = L + \left(\frac{\frac{N}{2} - F}{f}\right)C \tag{13}$$

Standard deviation (STD) is a range of how far the signal varies from the mean and average deviation. This variance value can be gain by the average of the squaring different each of the deviations. The variance represents the power of this fluctuation. In equation form, the standard deviation (14) and variance (15) is calculated:

$$s = \sqrt{\frac{1}{N-1}\sum\nolimits_{i=0}^{N-1}(x_1 - \mu)^2} \tag{14}$$

$$s^2 = \frac{1}{N-1}\sum\nolimits_{i=0}^{N-1}(x_1 - \mu)^2 \tag{15}$$

Skewness is a measurement of data has outliers which means has symmetry. Identify the signal whether the symmetric to the left or to right of the center point. The kurtosis data can be identify based on the normal distribution of right tailed or left tailed. Below is the mathematical equation:

$$Skewness = \frac{\frac{1}{N}\sum_{i=1}^{N}(x_i - \overline{x})^3}{\left(\frac{1}{N}\sum_{i=1}^{N}(x_i - \overline{x})^2\right)^{\frac{3}{2}}} \tag{16}$$

$$Kurtosis = \frac{\frac{1}{N}\sum_{i=1}^{N}(x_i - \overline{x})^4}{\left(\frac{1}{N}\sum_{i=1}^{N}(x_i - \overline{x})^2\right)^{2}} \tag{17}$$

4. Classification method

In order to complete this project, the construction of last part is handled in MATLAB software. The coding of classification PQ disturbance is created, and result is recorded. Simulation of SVM method is done to gain the accuracy of classification.

SVM classifier start with load a big number of PQ disturbance sample, i.e. 100. Which randomly consist of 9 types of statistical pattern. All data is randomize using for loop. After record the whole randomized data, 2 stages of SVM algorithm will be perform. Step first for SVM is performing training samples which based on flowchart been initialized at 50%. In order to produce good generalization performance function of Gaussian or polynomial can be chosen in training model. Then fitness function will be calculated referring on the training data Therefore, the remaining data sample are used as testing process. The training result will be model for testing process. Then the accuracy is evaluated whether reach satisfaction of this project or far from reaching desired accuracy.

Target value will be compared with testing. Should be the classification success approaching 100% to reach high accuracy level. If the classification evaluate is effective enough, thus process will be start again with randomized data.

Figure 8 above is created at editor window. The training data percentage is select for 50%. In randomize data code loop for is used. For loop function to executing a specific statement repeatedly until some false condition is met. In this application, it's for the purpose of different type of PQ disturbance. The Data is the original statistical feature of each sample. Then all the statistical data is identified which class each. It is arranged into their type of disturbance and statistical features is assemble randomly.

Figure 9 above is coding of training classification of PQ disturbances. X interval is the input of predictor data which each column will be observe while Y interval is target value. At line 40, there will be non-linear type of default value for the two-class learning that separate the data with the hyperplane. It is important to choose type of kernel function in order to accomplish success training. RBF is used for one type training while gaussian or polynomial can be used for two type class or multiple. Polynomial is chosen due to disorder symmetry. The 'class names' to distinguishes the different type data which represent PQ disturbances.

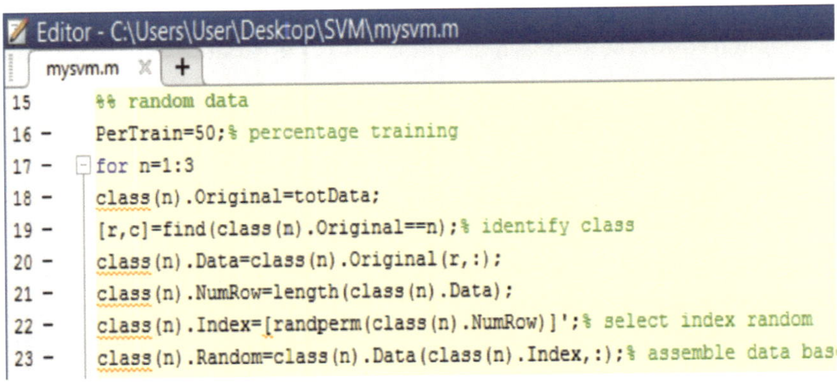

Figure 8.
Coding of randomized data of classification using SVM in MATLAB software.

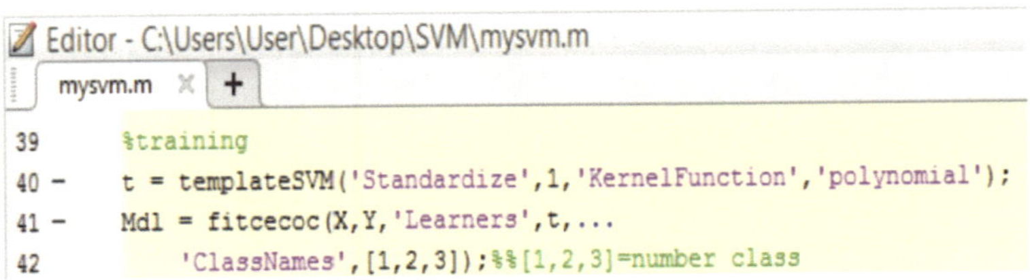

Figure 9.
Training coding of SVM in MATLAB software.

5. Conclusion

Further use of this coding will produce the classification result from the use of ST and SVM in the LabVIEW tools. This classification is significant, particularly in sending the electricity-based problem which originated from PQ disturbances. The LabVIEW function which assemble the ST, SVM and interfacing system in one platform could be a beneficial for electrical engineer conduct the problem shooting step. The capability of LabVIEW to integrate with different type algorithm prove to be the strength to materialize its function as classification tools. The introduction of this classification technique will minimize the time taken to pinpoint problem related to PQ disturbance. The more advance feature of this tools could be extended towards cloud computing function, Internet of Thing (IOT) application as well as wireless classification process.

Acknowledgements

We are grateful to the editors and anonymous reviewers for their constructive comments and valuable suggestions that helped to improve the quality of this manuscript significantly.

Appendices and nomenclature

AI	artificial intelligent
CWT	continuous wavelet transform

ECOC	error correction output code
EEMD	empirical embedded mode decomposition
GRNN	general regression neural network
HHT	Hilbert Huang transform
HT	Hilbert Huang
IOT	Internet of Thing
LMNN	Levenberg Marquardt neural network
NI	national instruments
OVO	one versus one
OVA	one versus all
PNN	probability neural network
PQ	power quality
MRA	multiresolution analysis
RMS	root-mean-square
RBF	radial basis function
RBFNN	radial basic function neural network
ST	S-transform
STD	standard deviation
STFT	short time fast transform
SVM	support vector machine
THD	total harmonic distortion
WT	wavelet transform

Author details

Ahmad Farid Abidin* and Mohd Abdul Talib Mat Yusoh
Faculty of Electrical Engineering, Universiti Teknologi MARA, Shah Alam, Selangor, Malaysia

*Address all correspondence to: ahmad924@uitm.edu.my

References

[1] Mazlumi K. Power Quality Monitoring. In: *Power Quality Monitoring, Analysis and Enhancement*. Rijeka: InTech, p. Ch. 1.

[2] Zobaa AF, Aleem SHEA, Balci ME. Introductory Chapter: Power System Harmonics—Analysis, Effects, and Mitigation Solutions for Power Quality Improvement. In: Zobaa A, Aleem SHEA, Balci ME (eds) *Power System Harmonics - Analysis, Effects and Mitigation Solutions for Power Quality Improvement*. Rijeka: InTech, p. Ch. 1.

[3] Polycarpou A. Power Quality and Voltage Sag Indices in Electrical Power Systems. In: Rey GR, Muneta LM (eds) *Electrical Generation and Distribution Systems and Power Quality Disturbances*. Rijeka: InTech, p. Ch. 6.

[4] Santos SR dos, Peres R, Cano WFR, et al. Simulation and Optimization of Electrical Insulation in Power Quality Monitoring Sensors Applied in the Medium-Voltage. In: Sánchez RA (ed) *Simulation and Modelling of Electrical Insulation Weaknesses in Electrical Equipment*. Rijeka: InTech, p. Ch. 7.

[5] Power I, Society E. *IEEE Guide for Application of Power Electronics for Power Quality Improvement on Distribution Systems Rated 1 kV Through 38 kV*. 2012. Epub ahead of print 2012. DOI: 10.1109/IEEESTD.2012.6190701.

[6] A. Rodriguez, J.E. Ruiz, J. Aguado, J.J. Lopez, F.I. Martin FM. Classification of Power Quality Disturbances using S-transform and Artificial Neural Networks. 2011 IEEE Int Conf Power Eng Energy Electr Drives 2011; 1: 608–615.

[7] Science C, Rourkela T. Feature Detection using Manish Bansal Feature Detection using S-Transform.

[8] Ramesh Babu N, Jagan Mohan B. Fault classification in power systems using EMD and SVM. *Ain Shams Eng J*. Epub ahead of print 2015. DOI: 10.1016/j.asej.2015.08.005.

[9] Jayasree T, Devaraj D, Sukanesh R. Power quality disturbance classification using Hilbert transform and RBF networks. Neurocomputing 2010; 73: 1451–1456.

[10] De Yong D, Bhowmik S, Magnago F. An effective power quality classifier using wavelet transform and support vector machines. Expert Syst Appl 2015; 42: 6075–6081.

[11] Jurado F, Saenz JR. Comparison between discrete STFT and wavelets for the analysis of power quality events. Electr Power Syst Res 2002; 62: 183–190.

[12] Fais M, Ghani A, Abidin AF, et al. Real Time Detection and Classification of Single and Multiple Power Quality Disturbance Based on Embedded S-Transform Algorithm in Labview. In: *Proceedings of the 2017International Symposium on Industrial Engineering and Operations Management (IEOM)*. 2017, pp. 280–289.

[13] Brazil USP, Kagan N, Ordoñez G. Automatic Power Quality Disturbance Classification Using Wavelet,. 2009; 8–11.

[14] Al-Aboosi YY, Sha'ameri AZ. Improved signal de-noising in underwater acoustic noise using S-transform: A performance evaluation and comparison with the wavelet transform. *J Ocean* Eng Sci 2017; 000: 1–14.

[15] Mahela OP, Shaik AG, Gupta N. A critical review of detection and classification of power quality events. Renew Sustain Energy Rev 2015; 41: 495–505.

[16] He Z, Ji Y. S-Transform Based Novel Indices for Power Quality Disturbances. In: Zobaa A, Canteli MM, Bansal R (eds)

Power Quality Monitoring, Analysis and Enhancement. Rijeka: InTech, p. Ch. 10.

[17] Lin WM, Wu CH, Lin CH, et al. Detection and classification of multiple power-quality disturbances with wavelet multiclass SVM. IEEE Trans Power Deliv 2008; 23: 2575–2582.

[18] Chen ZM, Li MS, Ji TY, et al. Detection and Classification of Power Quality Disturbances Using Probabilistic Neural Network. *2016 Int Jt Conf Neural Networks* 2016; 1277–1282.

[19] Ni YQ, Li M. Wind pressure data reconstruction using neural network techniques: A comparison between BPNN and GRNN. Measurement 2016; 88: 468–476.

[20] Fung CCFCC, Iyer V, Brown W, et al. Comparing the Performance of Different Neural Networks Architectures for the Prediction of Mineral Prospectivity. 2005 Int Conf Mach Learn Cybern 2005; 1: 18–21.

[21] Schwenker F, Kestler HA, Palm G. Three learning phases for radial-basis-function networks. Neural Networks 2001; 14: 439–458.

[22] Pahuja P, Chandra P, Scholar MT, et al. Power Quality Monitoring using LabView.

[23] El-Leathey L-A. Energy Management System Designed for the Interconnected or Islanded Operation of a Microgrid Using LabVIEW Software. In: Nayeripour M, Waffenschmidt E, Kheshti M (eds) *Smart Microgrids*. Rijeka: InTech, p. Ch. 3.

[24] Übeyli ED. Doppler ultrasound signals analysis using multiclass support vector machines with error correcting output codes. Expert Syst Appl 2007; 33: 725–733.

[25] Übeyli ED. Multiclass support vector machines for diagnosis of erythemato-squamous diseases. Expert Syst Appl 2008; 35: 1733–1740.

[26] Bosnic JA, Petrovic G, Putnik A, et al. Power Quality Disturbance Classification Based on Wavelet Transform and Support Vector Machine. *Proc 11th Int Conf Smolenice,* Slovakia 2017; 9–13.

[27] Zhu H, Liu X, Lu R, et al. Efficient and Privacy-Preserving Online Medical Prediagnosis Framework Using Nonlinear SVM. IEEE J Biomed Heal Informatics 2017; 21: 838–850.

[28] Li J, Teng Z, Tang Q, et al. Detection and Classification of Power Quality Disturbances Using Curvelet Transform and Support Vector Machines. Int Conf Inf Commun Embed Syst (ICICES2016) 2016; 1–11.

[29] Xiao-feng L, Xue-ying Z, Ji-kang D. Speech Recognition Based on Support Vector Machine and Error Correcting Output Codes. *2010* First Int Conf Pervasive Comput Signal Process Appl 2010; 336–339.

Chapter 9

Digital System Design

Janani Rajaraman

Abstract

The main objective of this chapter is to study and design various combinational circuits like Verification of Boolean Expression, Multiplexer, Demultiplexer Circuits, Code Converters circuits using LabVIEW tools. This chapter will make the user more comfortable towards learning of Design of Digital Systems. The various types of Boolean Expressions like SOP and POS, Combinational circuits like Adder circuit (Half adder and full adder), Subtractor circuit (Half Subtractor, Full Subtractor), some code converters like Binary to Gray and Gray to Binary, BCD to Gray and Gray to BCD and also Sequential circuits with D flip flop is also being carried out using this LabVIEW.

Keywords: Combinational Circuits, Multiplexer, Code Converters, Adder and Subtractor Circuits

1. Introduction

The field of electronics are classified into two broad group namely analog electronics and digital electronics. Analog electronics deals with signals that are continuous with respect to time by nature such as any noise signal, any video streaming etc. and digital electronics deals with signals that are discontinuous or discrete with respect to time. The electronic amplifier such as op-amp circuit helps to amplify the continuous signals and such signals are termed as *analog signals* and the circuit used for such applications are called *analog circuits.* On the other hand, the discrete signals are fed as the input to the computer by electronic switches, which as two distinct values such as HIGH level and LOW level [1]. This discrete signals are further converted in to electronic signals with the help of suitable converters. Such discrete signals are called as *digital signals* and the electronic circuit used for such operation are termed as *digital circuits.*

2. Boolean Algebra

In a discrete signals the two distinct values such as HIGH and LOW has equivalent voltage levels such as 5 volts and 0 volts respectively. This two distinct levels are represented as value 1 and value 0 respectively. Any algebraic functions performed with respect to this discrete values are defined as *Boolean algebra* developed by George Boole. He also developed various suitable theorems associated with this boolean for manipulation and simplification. There are set of basic definitions which are assumed to be true which defines all the information about the system. The following are the basic definition used in boolean algebra.

IntechOpen

- ***NOT:*** The NOT of a variable is 1 if and only if, the variable itself is 0 and vice versa
- ***AND:*** The AND of two variables is 1 if and only if *both* the variables are 1.
- ***OR:*** The OR of two variables is 1 if *either (or both)* of the variables is 1.
- ***XOR:*** The Exclusive-OR of two variables is 1 if *either* of them but *not both* is 1

3. Combinational logic circuits

There are two types of circuits exists in the digital system, Combinational Logic Circuits and Sequential Circuits. A combinational logic circuit is a circuit where the output depends on the combination of present input state. The set of operations which these combinational circuits performs logically by a set of Boolean functions. A sequential logic circuit is a circuit where the output depends on the combination of present input state and past input or previous input values. The previous output values are stored in the memory elements.

A combinational circuit consist of variables for input and output, and basic logic gates to perform the boolean function. The output signals are generated according to the inputs as well as the logic circuits employed. Here both the input and outputs are binary values either 1 or 0. **Figure 1** shows the simple block diagram of combinational logic circuits with *n* input variables and *m* output variables. If there are *n* number of inputs to the circuit then 2^n possible combinations of input states but each combination can produce only one output state [2]. For instance if the combinational logic circuit has 2 inputs A and B then there can be 4 possible input states.

3.1 Design procedure

The following are the steps considered while designing a combinational logic circuits.

- The first step is the statement of the problem for which this combinational circuit need to be designed
- Definition of the input and output variables and the variable name for inputs and outputs.
- Formation and tabulation of truth table which describes the relationship between the input and output.
- The simplified boolean expression is obtained for the output variable with the help of minimization techniques or Karanaugh map.

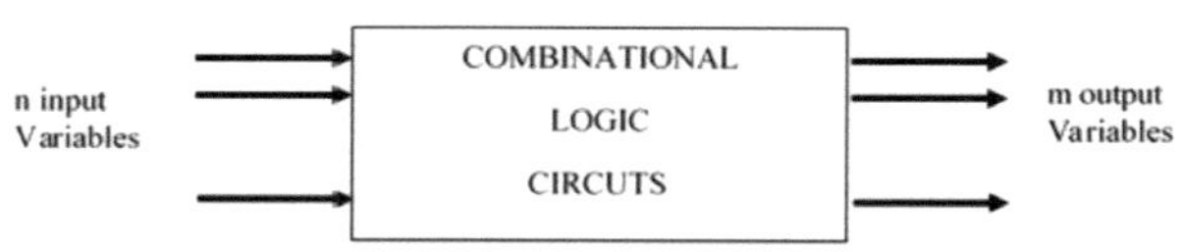

Figure 1.
Combinational Logic Circuits.

- The logical diagram using logic gates is realized for the simplified expression obtained in the previous step
- In practical design and real time implementation one should consider to use minimum number of gates.

3.2 Arithmetic circuits

3.2.1 Adder circuits

A combinational circuit that performs the addition of two bits is called *half-adder*. When the augend and addend numbers contain more significant digits, the carry obtained from the addition of two bits is added to the next higher order pair of significant bits. The combinational circuit that performs the addition of three bits is called a *full-adder*. The full adder can also be obtained by using two half adder circuits.

(i) Design of half-adders

A half adder is a combination logic circuit that uses two inputs (A and B) and two outputs (Sum S and Carry C). **Table 1** shows the truth table the various combinations of inputs and its corresponding outputs. The output Sum S and Carry C is obtained and the k-map is used to get the logical equation. The Boolean expressions are

$$Sum\ S = A \bigoplus B \tag{1}$$

$$Carry\ C = AB \tag{2}$$

Figure 2 shows the design and implementation of half adder circuit in LabVIEW environment, where the front panel that two inputs Input A and Input B, the outputs are Sum and Carry [3]. The block diagram in LabVIEW environment shows the logic gate implementation for the above obtained expression.

(ii) Design of full-adders

A full adder is a combination logic circuit that uses three inputs (A, B and C_{in}) and two outputs (Sum S and Carry C). **Table 2** shows the truth table the various combinations of inputs and its corresponding outputs. The output Sum S and Carry C is obtained and the k-map is used to get the logical equation.

$$Sum\ S = A \bigoplus B \bigoplus C_{in} \tag{3}$$

Input Variables		*Output Variables*	
A	**B**	**Sum S**	**Carry C**
0	0	0	0
0	1	1	0
1	0	1	0
1	1	0	1

Table 1.
Truth Table – Half Adder.

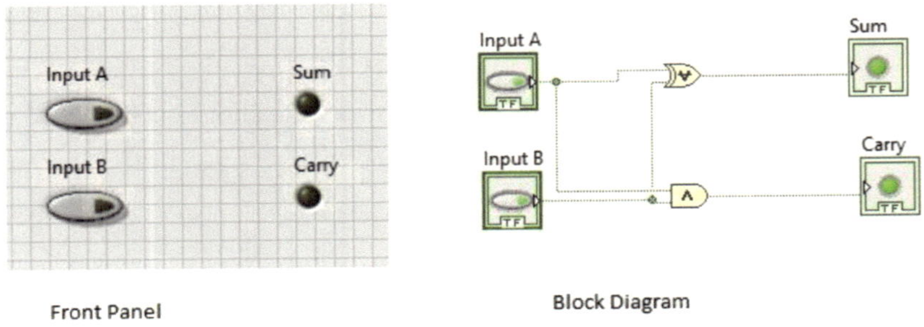

Figure 2.
Logic Diagram – Half Adder.

Input Variables			*Output Variables*	
A	**B**	**C_{in}**	**Sum S**	**Carry C**
0	0	0	0	0
0	0	1	1	0
0	1	0	1	0
0	1	1	0	1
1	0	0	1	0
1	0	1	0	1
1	1	0	0	1
1	1	1	1	1

Table 2.
Truth Table – Full Adder.

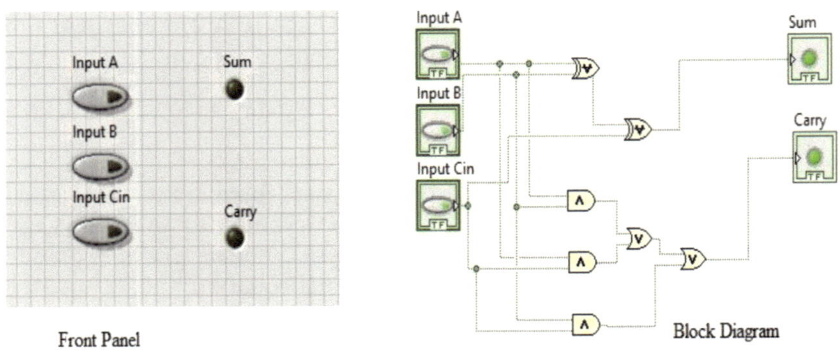

Figure 3.
Logic Diagram – Full Adder.

$$Carry\ C = AB + BC_{in} + C_{in}A \tag{4}$$

Figure 3 shows the design and implementation of full adder circuit in LabVIEW environment, where the front panel that two inputs Input A, Input B, and Input C_{in}, the outputs are Sum and Carry. The block diagram in LabVIEW environment shows the logic gate implementation for the above obtained expression.

3.2.2 Subtractor circuits

A combinational circuit that performs the difference of two bits is called *half-subtractor*. When the first input (minuend) is 0 and the second input(subtrahend) is 1 then there exists a output variable as *Borrow*. The combinational circuit that determines the difference of three bits is called a *full-subtractor*.

(i) Design of half-subtractor

A half subtractor is a combination logic circuit that uses two inputs (A and B) and two outputs (Difference D and Borrow B). **Table 3** shows the truth table the various combinations of inputs and its corresponding outputs. The output Difference D and Borrow B is obtained and the k-map is used to get the logical equation. **Figure 4** shows the design and implementation of half subtractor circuit in LabVIEW environment, where the front panel that two inputs Input A and Input B, the outputs are Difference and Borrow

$$Difference = A \oplus B \tag{5}$$

$$Borrow = \overline{A}B \tag{6}$$

(ii) Design of Full-subtractor

A full subtractor is a combination logic circuit that uses three inputs (A, B and B_{in}) and two outputs (Difference D and Borrow B). **Table 4** shows the truth table the various combinations of inputs and its corresponding outputs. The output difference D and Borrow B is obtained and the k-map is used to get the logical equation

$$Difference\ D = A \oplus B \oplus B_{in} \tag{7}$$

$$Borrow\ B = \overline{A}B + A \oplus B \tag{8}$$

Input Variables		***Output Variables***	
A	**B**	**Difference D**	**Borrow B**
0	0	0	0
0	1	1	1
1	0	1	0
1	1	0	0

Table 3.
Truth Table – Half Subtractor.

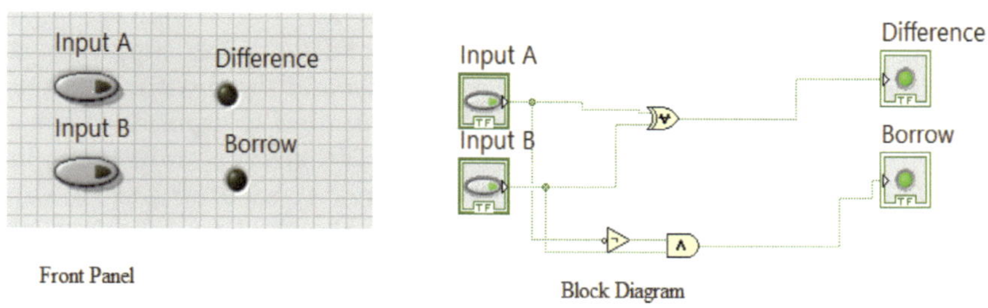

Figure 4.
Logic Diagram – Half Subtractor.

Input Variables			*Output Variables*	
A	**B**	B_{in}	**Difference D**	**Borrow B**
0	0	0	0	0
0	0	1	1	1
0	1	0	1	1
0	1	1	0	1
1	0	0	1	0
1	0	1	0	0
1	1	0	0	0
1	1	1	1	1

Table 4.
Truth Table – Full Subtractor.

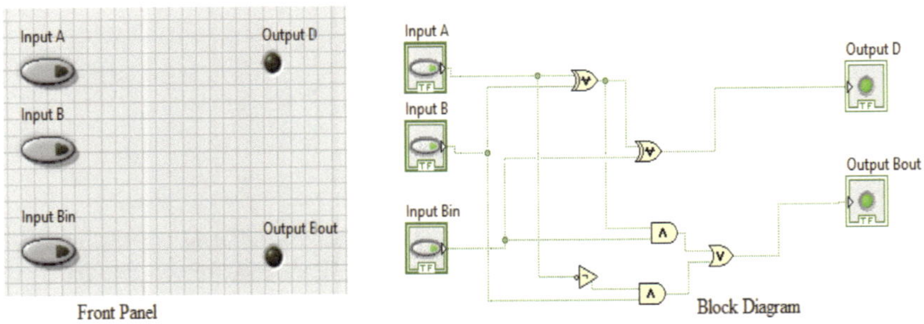

Figure 5.
Logic Diagram – Full Subtractor.

Figure 5 shows the design and implementation of full subtractor circuit in LabVIEW environment, where the front panel that two inputs Input A, Input B, and Input B_{in}, the outputs are Difference D and Borrow Ouptut. The block diagram in LabVIEW environment shows the logic gate implementation for the above obtained expression.

3.3 Multiplexer and demultiplexer

The most important form of a combinational circuit and which is widely used in the field of communication is *Multiplexer and Demultiplexer Circuits.* The multiplexer circuit is used to transfer large number of channels carrying information to a smaller number of channels. Such circuit used to transmit digital data or binary information is called as *data selector or digital multiplexer.* In this data selector, the input line is selected according to the combination of select lines, suppose if there exists 2^n input line then the number of select line is *n* and there will be *only one* output line. For example in a 4x1 multiplexer, the number of input lines is 4 (2^2) which shows there exists of *2* select lines. In these multiplexer circuits the inputs are named as *I0, I1, I2 and I3* and the two select lines are named as *S0* and *S1.* **Table 5** shows the various combinations of select line and corresponding input line is selected and obtained as output Y [4]. The boolean expression for the output Y is given below. **Figure 6** shows the Front panel and Block Diagram of 4x1 multiplexer. This design can be extended for higher versions like 8x1 and 16x1 types of multiplexer.

Selection Inputs		Input Channels				Output
S1	**S0**	**I0**	**I1**	**I2**	**I3**	**Y**
0	0	0	X	X	X	0
0	0	1	X	X	X	1
0	1	X	0	X	X	0
0	1	X	1	X	X	1
1	0	0	X	0	X	0
1	0	1	X	1	X	1
1	1	0	X	X	0	0
1	1	1	X	X	1	1

Table 5.
*Truth Table – 4*1 Multiplexer.*

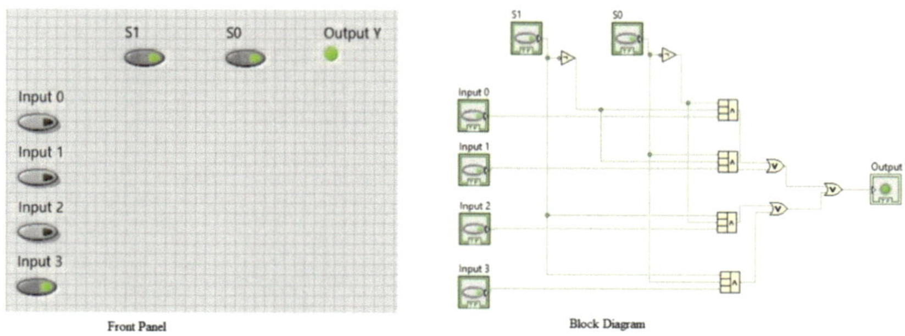

Figure 6.
*Logic Diagram – 4*1 Multiplexer.*

The Boolean expressions are

$$Output\ Y = \overline{S_0 S_1} I_0 + \overline{S_0} S_1 I_1 + S_0 \overline{S_1} I_2 + S_0 S_1 I_3 \tag{9}$$

The term demultiplex is just a opposite way of multiplexer, here in this combinational circuit there are one input channel and distributes the data over several channels. Therefore if the number of input channel is *1* then the number of output will be 2^n output channels. The combination of select lines control the output channel through which the input data must be transmitted. **Table 6** gives the truth table for 1-to-8 demultiplexer, the front panel and block diagram for 1-to-8 demultiplexer is shown in **Figure 7**.

The selection input line S0, S1, S2 are activated according to the bit combination for each output as given in Eq. (10) to Eq. (17). For instance, if the selection input combination is 010, the input *I* is transmitted to Y2. The Boolean expressions for each output line is given below

$$Y0 = \overline{S_2 S_1 S_0} I \tag{10}$$

$$Y1 = \overline{S_2 S_1} S_0 I \tag{11}$$

$$Y2 = \overline{S_2} S_1 \overline{S_0} I \tag{12}$$

$$Y3 = \overline{S_2} S_1 S_1 I \tag{13}$$

$$Y4 = S_2 \overline{S_1 S_0} I \tag{14}$$

Selection Inputs			Output Channels							
S2	S1	S0	Y0	Y1	Y2	Y3	Y4	Y5	Y6	Y7
0	0	0	Y0 = I	0	0	0	0	0	0	0
0	0	1	0	Y1 = I	0	0	0	0	0	0
0	1	0	0	0	Y2 = I	0	0	0	0	0
0	1	1	0	0	0	Y3 = I	0	0	0	0
1	0	0	0	0	0	0	Y4 = I	0	0	0
1	0	1	0	0	0	0	0	Y5 = I	0	0
1	1	0	0	0	0	0	0	0	Y6 = I	0
1	1	1	0	0	0	0	0	0	0	Y7 = I

Table 6.
*Truth Table – 1*8 Demultiplexer.*

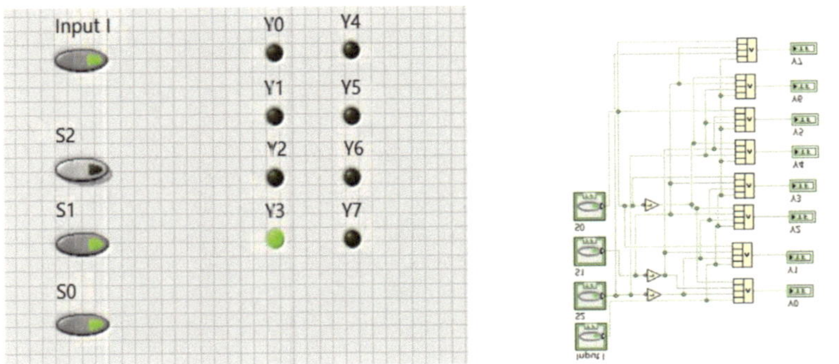

Figure 7.
*Logic Diagram – 1*8 Demultiplexer.*

$$Y5 = S_2\overline{S_1}S_0I \quad (15)$$

$$Y6 = S_2S_1\overline{S_0}I \quad (16)$$

$$Y7 = S_2S_1S_0I \quad (17)$$

3.4 Code converters

For the same discrete elements of information, there are several different codes available, resulting in the use of different codes for different digital systems. It's sometimes necessary to connect two digital blocks that use different coding systems. Hence a conversion digital circuit is designed and implemented between two digital systems to use information of one digital system to another. The input lines must provide the bit combinations of elements as designed by binary code A and the output is generated by the bit combinations of code B. This code converters circuit consisting of logic gates to perform this transformation operations. Some of the few code conversion techniques are discussed below.

3.4.1 Binary to gray code converters

Table 7 shoes the conversion of 4-bit binary code to its equivalent gray code values. The 4 bit binary code input is defined as B_0, B_1, B_2, B_3 and corresponding

DOI: http://dx.doi.org/10.5772/intechopen.97611

B_0	B_1	B_2	B_3	G_0	G_1	G_2	G_3
0	0	0	0	0	0	0	0
0	0	0	1	0	0	0	1
0	0	1	0	0	0	1	1
0	0	1	1	0	0	1	0
0	1	0	0	0	1	1	0
0	1	0	1	0	1	1	1
0	1	1	0	0	1	0	1
0	1	1	1	0	1	0	0
1	0	0	0	1	1	0	0
1	0	0	1	1	1	0	1
1	0	1	0	1	1	1	1
1	0	1	1	1	1	1	0
1	1	0	0	1	0	1	0
1	1	0	1	1	0	1	1
1	1	1	0	1	0	0	1
1	1	1	1	1	0	0	0

Table 7.
Truth Table – Binary to Gray Converter.

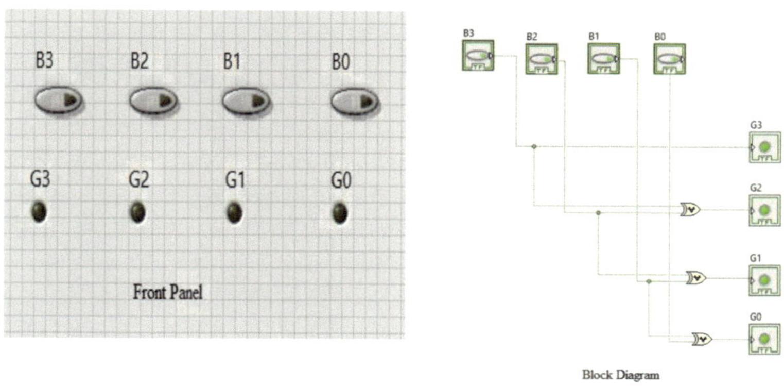

Figure 8.
Logic Diagram – Binary to Gray Code Converter.

output 4-bit gray is defined as G_0, G_2, G_3, and G_4 as shown in **Figure 8**. The corresponding boolean expression for binary to gray code conversion is given below

$$G_0 = B_0 \tag{18}$$

$$G_1 = B_0 \bigoplus B_1 \tag{19}$$

$$G_2 = B_2 \bigoplus B_1 \tag{20}$$

$$G_3 = B_3 \bigoplus B_2 \tag{21}$$

3.4.2 Gray to binary code converters

Gray code is also called as Reflected Binary Code (RBC), Reflected Binary (RB) or Gray code, Cyclic Code, is defined as an ordering of the binary number system such that each incremental value can only differ by one bit. The main objective in this code converter is that while traversing from one step to another step, one bit in the code group changes as in **Figure 9**. This gray code is not applicable for arithmetic operations, but it is applicable in analog to digital converters, as well as error correction techniques in digital communications (**Table 8**).

$$B_3 = G_3 \tag{22}$$

$$B_2 = G_2 \oplus G_3 \tag{23}$$

$$B_1 = G_1 \oplus G_2 \oplus G_3 \tag{24}$$

$$B_0 = G_1 \oplus G_2 \oplus G_3 \oplus G_0 \tag{25}$$

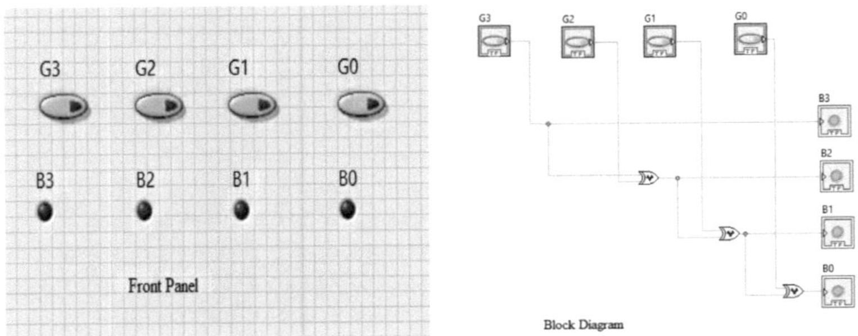

Figure 9.
Logic Diagram – Gray to Binary Code Converter.

G_3	G_2	G_1	G_0	B_3	B_2	B_1	B_0
0	0	0	0	0	0	0	0
0	0	0	1	0	0	0	1
0	0	1	1	0	0	1	0
0	0	1	0	0	0	1	1
0	1	1	0	0	1	0	0
0	1	1	1	0	1	0	1
0	1	0	1	0	1	1	0
0	1	0	0	0	1	1	1
1	1	0	0	1	0	0	0
1	1	0	1	1	0	0	1
1	1	1	1	1	0	1	0
1	1	1	0	1	0	1	1
1	0	1	0	1	1	0	0
1	0	1	1	1	1	0	1
1	0	0	1	1	1	1	0
1	0	0	0	1	1	1	1

Table 8.
Truth Table – Gray to Binary Code Converter.

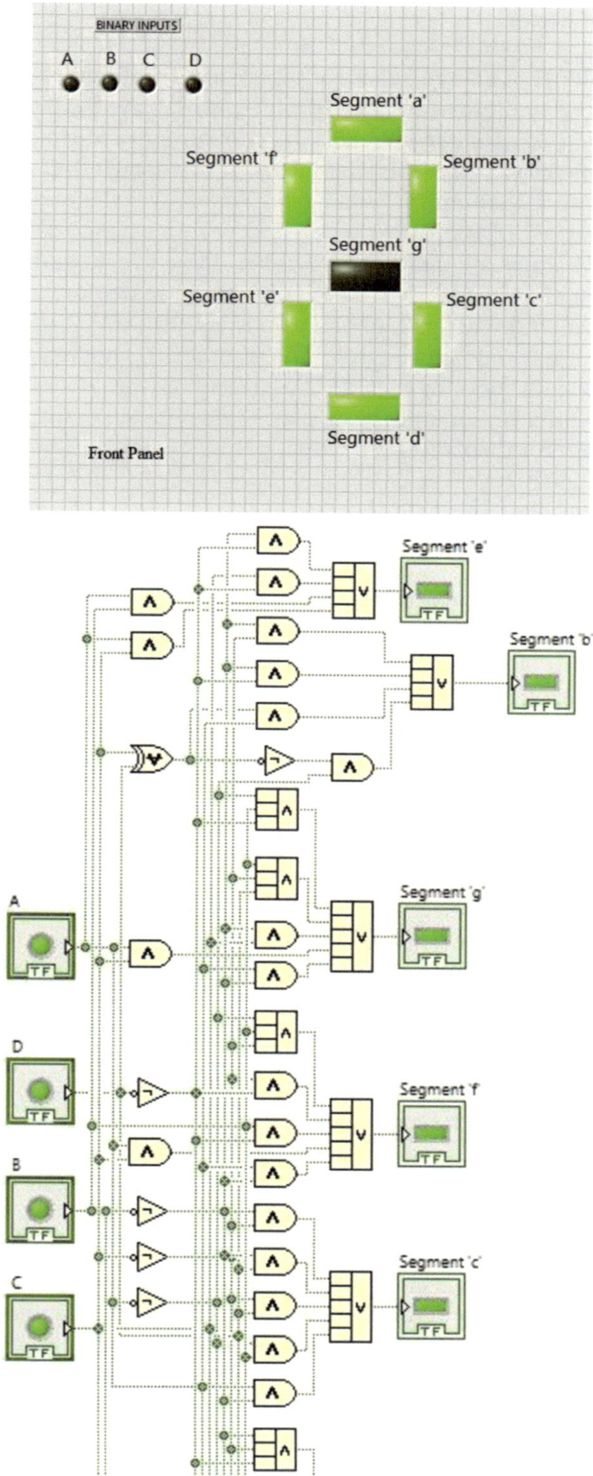

Figure 10.
Logic Diagram – Seven Segment Decoder.

3.4.3 Seven segment decoder

A digital decoder IC is a device that converts one digital format into another, and one of the most commonly-used device for doing this is the binary-coded decimal (BCD) to 7-segment display decoder. The 7-segment light emitting diode (LED) provides a convenient way of displaying information or digital data in the form of numbers, letters and alphanumeric characters. Typically, 7-segment displays consist of seven same coloured LEDs (called segments) within a single display package. In order to display the correct character or number, the correct combination of LED segments has to be illuminated. This LabVIEW program demonstrates the illumination of each segment by displaying hex values (0000 through FFFF) in decimal form from 0 through 9 and A through F. The standard 7-segment LED display has eight input connections, one for each LED segment and one that acts as a common terminal or connection for all internal display segments. Some displays also have an additional input pin for displaying a decimal point.

3.4.4 Types of digital display

There are two important types of 7-segment LED displays, namely, common cathode and common anode. In Common cathode display (CCD) display, all cathode connections of the LEDs are joined together to a low logic or ground or 0 [5]. The individual segment is illuminated by the application of high logic or + Vcc or 1 to the individual anode terminal. In Common anode display (CAD) In a CAD, all anode connections of the LEDs are joined together to a high logic or + Vcc and individual segments are illuminated by connecting individual cathode terminals to low logic or ground as shown in **Figure 10**. The boolean expressions of the outputs.

$$a = \overline{A}BD + A\overline{BC} + \overline{BD} + AC + A\overline{D} + BC \tag{26}$$

$$b = \overline{A}C \bigoplus D + A(C \bigoplus D) + \overline{BC} + \overline{BD} \tag{27}$$

$$c = \overline{BC} + \overline{B}D + \overline{C}D + \overline{A}B + A\overline{B} \tag{28}$$

$$\mathrm{d} = \overline{BCD} + \overline{B}DC + \overline{A}C\overline{D} + B\overline{C}D + A\overline{C} + AB\overline{D} \tag{29}$$

$$e = \overline{BD} + C\overline{D} + AB + AC \tag{30}$$

$$f = \overline{A}B\overline{C} + \overline{CD} + B\overline{D} + A\overline{B} + AC \tag{31}$$

$$g = \overline{A}B\overline{D} + B\overline{C}D + A\overline{B} + AC + C\overline{B} \tag{32}$$

4. Conclusion

This chapter brings an overview of design of combinational logic circuits in LabVIEW. This LabVIEW programming tool is a graphical representation tool which helps the designer to simplify the design work. This tool can be further extended for designing sequential circuits as well as PLA and PAL logic design.

Acknowledgements

The author would like to thank the Department of Electronics and Instrumentation Engineering of Sri Chandrasekharendra Saraswathi Viswa Mahavidyalaya, Enathur, Kanchipuram, India, for providing online library facility as well as for permitting to use the lab environment.

DOI: http://dx.doi.org/10.5772/intechopen.97611

Author details

Janani Rajaraman
Department of Electronics and Instrumentation Engineering,
Sri Chandrasekharendra Saraswathi Viswa Mahavidyalaya, Enathur,
Kanchipuram, Tamilnadu, India

*Address all correspondence to: janani.rajaraman@kanchiuniv.ac.in

References

[1] Holdsworth, Brian and Clive Woods, *Digital Logic Design*, Elsevier, 2002.

[2] Mano, M. Morris and Charles R. Kime, *Logic and Computer design fundamentals,* Prentice-Hall, Inc, 1997

[3] Dietmeyer, Donald Leo, *Logic Design of Digital Systems*, 1978.

[4] Friedman, Arthur D, *Logical Design of Digital Systems*, 1977.

[5] Rajaraman Janani and K. Saraswathi, *A LabVIEW based Monitoring and Controlling of Various Process Variables",* International Journal of Advanced Research in Electrical, Electronics and Instrumentation Engineering, Vol.3, 2014.